PALEOENVIRONMENTS IN THE NAMIB DESERT

THE LOWER TUMAS BASIN IN THE LATE CENOZOIC

by

M. Justin K. Wilkinson

Lockheed Engineering & Sciences Company

UNIVERSITY OF CHICAGO GEOGRAPHY RESEARCH PAPER NO. 231

The University of Chicago
Committee on Geographical Studies
1990

Library of Congress Cataloging-in-Publication Data

Wilkinson, M. Justin, 1946-
Paleoenvironments in the Namib Desert : the lower Tumas Basin in the Late Cenozoic / by M. Justin Wilkinson.
p. cm. -- (University of Chicago geography research paper ; no. 231)
Includes bibliographical references (p.) and index.
ISBN 0-89065-138-8 (pbk.) :
1. Paleoecology--Namibia--Tumas River Watershed. 2. Geology, Stratigraphic--Cenozoic. 3. Geology--Namibia--Tumas River Watershed. 4. Geology--Namibia--Namib Desert. I. Title. II. Series.
QE720.W55 1990
560'.178'096881--dc20 90-11272

Geography Research Papers are available from:

The University of Chicago
Committee on Geographical Studies
5828 South University Avenue
Chicago, Illinois, USA 60637-1583

To my wife, Sally, and daughter, Kate,
and to my mother and father

Contents

Figures

Tables

Preface

This monograph describes the geomorphic evolution of the Tumas River basin, one of the larger streams that rise below the Great Escarpment in the Central Namib Desert. The main thrust of the study has been an attempt to assess the paleoenvironmental import of various geomorphological, geological, and pedological phenomena of the Tumas drainage basin.

The status of basic knowledge with respect to the landscape, soil, and sediments in the Central Namib is so rudimentary that the study seemed appropriate in terms of description alone, of what may be a typical Namib Desert drainage basin. Such description may indeed have wider applicability in understanding the nature of fluvial activity in terminal, aggrading inland basins, since these are little understood in geomorphology or geology.

The chronology of climatic events documented here has broadened the picture of past west coast environments. Causes for the progression have by no means been provided. If anything, controversy on the matter is sharper. This is far from disappointing because thought on paleoclimates of the subcontinent appears to be enjoying a resurgence. The east-west climatic gradient of the Namib recurs constantly as a theme in this study, adding complexity to the interpretation of past landscapes. The challenge of reconciling formative geomorphic environments belied by the existence of neighboring but different pedologic landscapes remains.

Arid river termini are poorly understood as settings for ephemeral stream deposition. Since appreciation of the sedimentary bodies in the Tumas basin is a primary aim of this study, I have attempted to model those geomorphic components of the arid alluvial plain that determine depositional environment, and then to assess the meaning of changes in such environments. The connection between geomorphic and sedimentary environments has much theoretical interest when the panoply of internal and external determinants of fluvial sedimentation systems is considered. Of all these determinants, that of climate in particular has been a primary concern in this study.

Appreciation of the Tumas sediments in regional context has been possible only because of the existence of a regional stratigraphic sequence, the erection of which is a tribute to the dedication of past researchers. Correlation of the study area stratigraphy with wider sequences appears more feasible than originally expected. Recent understanding of the eustatic history of southern Africa's west coast has enhanced interpretation and dating of Tertiary events in the Namib.

The study was undertaken under the guidance of Karl W. Butzer while I was based at the University of Chicago. His work inspired me to examine the practice of geomorphology with an increased awareness of sediments, soils, and climatic controls. For this I owe a debt of gratitude.

Prospecting activities by the Anglo American Corporation presented me with an introduction to the Central Namib Desert. I would like to express my appreciation to the corporation for logistical support provided for three extended field seasons and for numbers of detailed maps and aerial photographs that were generously made available. Sediment and other analyses were undertaken in the Geography Department's Paleoecology Laboratory at the University of Chicago.

I thank Edmund Antrobus (Johannesburg) for initiating what has proven to be an unusual and varied research project in a unique part of the world. Early discussions with field geologists in the remote Husab field camp, especially with Brian Hambleton-Jones (Pretoria) and Phillip Woodhouse (Johannesburg), and with scientists at the Desert Ecology Research Unit (Gobabeb), were important in orienting a novice to the Namib Desert. Henk Gewald and Frans Wagener (Windhoek) took interest in the project. The latter assisted in obtaining specialized soil analyses. Michael Harrison (Johannesburg) provided stimulation to a jaded paleoclimatic palate. Ian Ralston (Johannesburg) and Al Levinson (Calgary) generously gave time to consider problems of the interpretation of uranium-series dates. Paul B. Moore (Chicago) introduced me with enthusiasm to X-ray diffraction techniques. Mario Barbafiera and Richard Lunz (Johannesburg) contributed respectively to discussions on local winds and operation of a statistical package.

I thank librarians, at the Universities of Chicago, Arizona and the Witwatersrand particularly, for consistent, specialized help in tracing obscure bibliographic material. I thank my colleagues at the University of the Witwatersrand for help in many ways, and especially—with members of the computer Centre—for their patience in refining my use of word processors. Wendy Job constructed many of the diagrams in masterful fashion. Members of the Graphics Centre produced numerous photographic prints. The Chief Director of the Department of Surveys and Mapping kindly made available several aerial photographs of the study area. The space image of the Central Namib was received and processed by the Satellite Remote Sensing Centre of the Council for Scientific and Industrial Research at Hartebeeshoek. Carol Saller performed the many functions of editor with great expertise, concern, and consideration.

I am especially pleased to be able to acknowledge the support of my wife, Sally Antrobus, and also that of our parents. Gary Rydout and the Andrew, Cox, Robb, and Solot families—friends in Tucson and Johannesburg—do not perhaps realize the importance of their help and generosity to the success of the project. Rodney and Heather Wilkinson (Cape Town), Renate Witt (Swakopmund) and Beatrice Sandelowsky (Windhoek) provided much-needed hospitality. I thank them all warmly.

I would like to express keen appreciation to the Department of Geography at the University of Chicago for supporting much of my graduate study, which was for me a period of exploration and great stimulation. I thank Chauncy D. Harris and Donald Collier for encouragement throughout the research period.

Chapter 1

INTRODUCTION

Losing the way . . . is the rule here . . . and the person who has crossed the [Naarip] plain without doing so rather plumes himself upon the feat.

—C. J. Andersson, *Lake Ngami*

. . . the [Naarip] plain is covered with false waggon-roads in every direction . . .

—F. Galton, *The Narrative of an Explorer in Tropical South Africa*

. . . the weary eye seems to range in vain over this howling wilderness [of the Naarip desert], in search of something worthy of attention.

—C. J. Andersson, *Notes on Travel in South Africa*

The Tumas River basin is a wide depression of low topography occupying much of the area between the great canyons of the Swakop and Kuiseb rivers in the Central Namib Desert (figure 1). The Tumas Flats and especially the lower Tumas valley are the focus of this study. The Tumas Flats until this century were commonly referred to as the Naarip[1] Desert or Naarip Plain, names that appear to have been lost entirely.

The study area is the rocky but flat lower section of the Tumas valley (rendered Tubas, Dumas, and Dupas on older maps and Domas by Stengel 1964), the largest of those streams that rise below the Escarpment of

---

1 Nienaber and Raper (1977:847) give variations Narip and Narriep; they comment that these may be prior forms of the well-known name Namib (also Naamib), which does not appear on early maps and documents. Despite earlier forms with known meanings (Nanib/!Nanib = "ridge," p. 854, and Narib = "congealed blood," p. 864) and popularly assigned meanings such as "desert," "emptiness," "wide flat country," and "God's country," these authors consider that the meaning of Namib is presently unknown. Namib forms the basis of the country name Namibia.

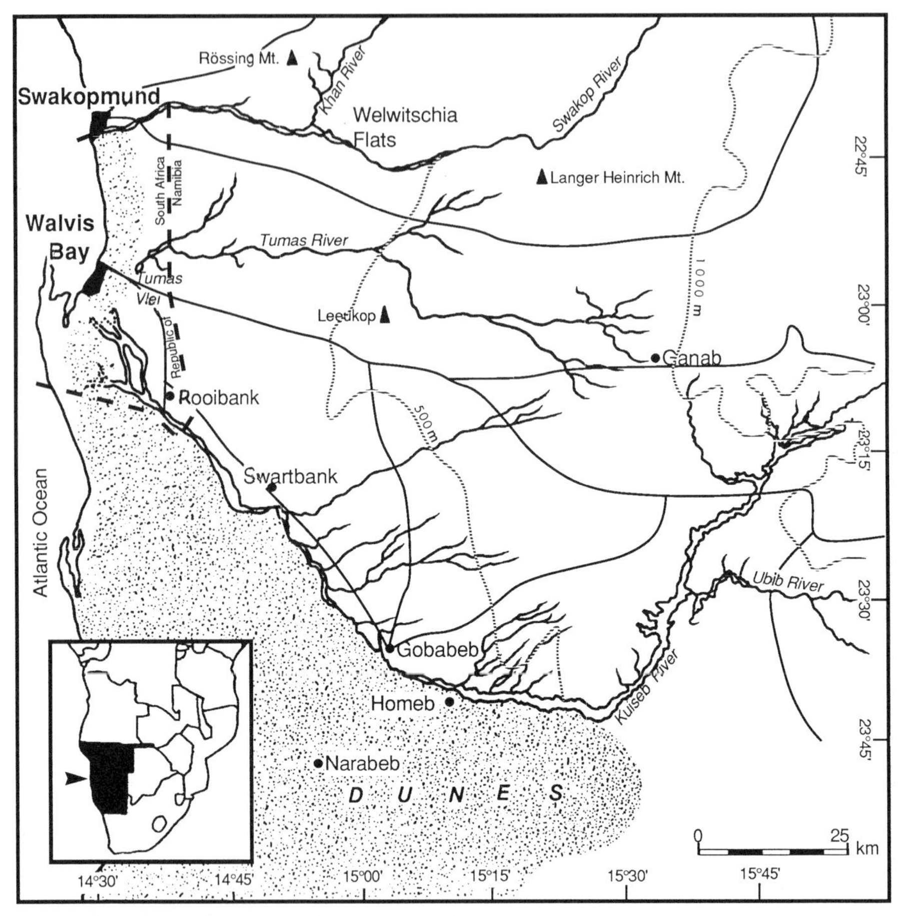

Fig. 1. Central Namib Desert and study area: physical and cultural features (500 m contour interval).

Namibia within the 100-150 km-wide west coast Namib Desert (Stengel 1964).

The core of the study area is the lower axial section of the Tumas drainage basin 10-60 km inland, where several geomorphic features are well developed and a large body of subsurface data is available from recent mineral exploration efforts. The Tumas River flows west at latitude 23°20'S to a terminal playa 10 km northeast of Walvis Bay.

The study area occupies a subdivision of the Namib Desert referred to by most writers as the Central Namib, defined by Spreitzer (1965-66) as stretching from the Ugab River and Brandberg massif in the north (approximately latitude 21°S) to the Kuiseb River in the south (approximately latitude 24°S). This area is set off geomorphically, botanically, and economically from the Skeleton Coast and Kaokoveld to the north, and from the Dune

Namib (Logan 1972), an area of high, dominantly linear dunes, which stretches southward 350 km.

Wellington (1955) considered the Tumas basin as part of the Damara or northern Namib. In the scientific literature "Central Namib" is the commonest term applied to the study area and its surroundings, however, and is used hereafter.

Several rationales exist for undertaking a geomorphic study of the lower Tumas basin. At the descriptive level, the investigation seems justified because the Central Namib Desert had been examined in a general way only until mineral exploration of surface and near-surface deposits yielded data in the mid-1970s.

The Central Namib has received scant attention, even in terms of geomorphic description, compared with the canyon zones of the major allochthonous Swakop and Kuiseb rivers, coastal features, and the Dune Namib to the south. Such description is of more than purely local interest because most of the area of the coastal platform is drained by watercourses such as the Tumas River that rise within the Namib Desert.

Since 1975 the Central Namib has become the center of uranium exploration by French, U.S., and South African mining corporations. The study area is now one of the most intensively investigated, with detailed geological and topographic mapping, aerial photography, and 10,000 m of surficial drilling data. A variety of geochemical and geophysical data and an extensive new pitting program in a core area of 30 km x 70 km have been made available by Anglo American Corporation of South Africa. This mass of information provides an unusual opportunity for research in a relatively uncharted area. Logistical support was given the writer for a total of nine months' field activity.

Another rationale stems from the relative importance of a historical geomorphological approach (*sensu* Butzer 1973) for the reconstruction of past landscapes in the Central Namib: other common sources of paleoenvironmental data, especially faunal and floral, seem to be lacking.

A further rationale is the need for paleoenvironmental data for the climatic province that constitutes the Namib Desert. Little is known on this head for the western parts of the subcontinent. In this respect the study area is ideally located in the hyperarid west coast desert core. In terms of morphogenetic theory of landform evolution, a climatic regime of stable temperatures, and more especially extreme precipitation conditions, should promote detection of oscillations away from present norms. Furthermore, not only does the Tumas River rise within the Namib Desert, but it is insulated from possible eustatic influences in the more recent past by outcropping bedrock bars across its mouth at the terminal playa. Isolation of climatic effects in the coastal desert are thus more securely based than might otherwise have been the case.

Finally, the study contributes to certain related aspects of arid-zone geomorphology, especially in the realms of desert sedimentary environments, stream behavior, and description of extensive gypsum accumulations and their interrelationships.

In general terms, the main theoretical background of the study comprises concepts of morphogenesis as expounded by Erhart (1967), modified to encompass regional variations. Butzer (1976a:140) has given one of the few explications in English of Erhart's writings: "In this interpretation of periodic landscape evolution, episodes of accentuated erosion lead to accelerated landform sculpture and net, long-term soil erosion, transport and deposition. Such morphodynamic episodes alternate with intervals of comparative stability and slow geomorphologic change, during which the rate of weathering equals or exceeds the rate of erosion."

These ideas, more current in European geomorphology, have proved fertile in southern Africa, where a series of studies in the last decade have documented many changes of geomorphic environment of the type that has been termed periodic. Morphogenetic phases, conceived as functioning systems, then provide a historical perspective when viewed chronologically.

The analysis of deposits in the Tumas basin is undertaken in terms of fluvial sedimentological concepts of Waltherian facies successions, facies models as formulated particularly by Miall (1977, 1978, 1984), and appreciation of the three-dimensional architecture of individual strata (Friend 1983; Miall 1985). Fluvial geomorphologic concepts of stream morphology and equilibrium states of longitudinal profiles provide the basis of analysis of later stream incision episodes. A succession of gypsum crust morphologies, paralleling in some ways those of caliche profiles, is identified. Examination of eolian features from remotely sensed images of the dune sea south of the study area contributes ideas on the timing of the most arid morphogenetic phases in the Central Namib.

After tangential Portuguese landings on the Namibian coast at Cape Cross (north of Swakopmund) and Elizabeth Point (Lüderitzbucht) in 1484 and 1486, European contact with the Central and southern Namib Desert was probably nonexistent until the late 1700s (Vedder 1938).

The late 1600s witnessed two unsuccessful voyages of discovery commissioned from Cape Town; minimal contact was made with local people in Namibia, and even such features as the mouths of the Kuiseb, Swakop, and Omaruru rivers were not discovered (Vedder 1938).

Pieter Pienaar, a hunter from the Cape Colony, arrived in Walvis Bay—one of only two natural harbors on the Namibian coast—by ship in

1792. He appears to have been the first European to have explored the Swakop River valley where it traverses the Namib Desert. Heinrich Schmelen, a missionary from the northern Cape Colony, reached the Kuiseb River and Walvis Bay (figure 1) in 1824 and 1825 by an overland route. Sir James Alexander described his journey by ox wagon from Cape Town to Walvis Bay, which he reached in 1837, documenting for the first time the rare event of the Kuiseb River "coming down in flood." Alexander's is the first mention of the "bare and extensive plains" (Alexander 1838: 63)—the subject of this study—which stretch north from the Kuiseb River canyon, and of "the range called Tumas, or the Mountains of the Wilderness,"[2] a low massif at the head of the Tumas River valley (Alexander 1838: 113).

These apparently featureless plains generally go unremarked today in descriptions of the Central Namib Desert, and the Tumas River is often omitted from maps. The shallow basin of the Tumas has been overshadowed in terms of geomorphic and geologic interest—the latter concerning orogenic modeling and uranium ore emplacement—by the major canyons and flanking badland topography of the Kuiseb and Swakop rivers to north and south, by the Dune Namib south of the Kuiseb River (figure 2), where some of the largest dunes in the world rise up, and by the more accessible coastal features.

North of the Swakop River lie the geological curiosities of the Erongo, Brandberg, and Spitzkoppe massifs, the largest uranium mine in the world at Rössing Mountain, and also the Welwitschia Flats, home of the well-known endemic desert plant *Welwitschia mirabilis*. All of these have focused attention away from the Tumas Flats on the south side of the Swakop River canyon (figure 1).

Economic activity has also skirted the lower Tumas River valley. Cattle ranching begins in the semiarid grasslands and thornscrub to the east and ascends the Great Escarpment onto the Khomas Hochland. To the west, the cool foggy desert coast has been the focus of whaling—especially by North American vessels as long ago as the 1700s—and more recently fishing and tourism, based at Walvis Bay and Swakopmund.

The north end of the Dune Namib, the Tumas Flats and the canyonlands of the Swakop, Khan, and Kuiseb rivers now form part of a nature re-

---

2 On the meaning of the word *Tumas*, Nienaber and Raper (1977:1067) have noted that "Alexander se vertaling ["Mountains of the Wilderness"] 'byna reg' is, dat dit beter te verklaar is met 'Die Berge van Verlatenheid' [Mountains of Desolation] of ' . . . Eensaamheid [Loneliness]'. . . . Die naam is dan te verbind met !u- soos in !u-eisa = eensaam. . . , of by !u = slap wees, verwelk, kragteloos. . . . Die uitgang -ma-s is . . . 'n versteende lokatiefformans."

Fig. 2. Part of the Central Namib Desert (34,000 km^2). 1. Lower Tumas River alluvial plain. 2. Tumas Vlei (playa). *3. Gawib Flats. 4. Welwitschia Flats. 5. Swakopmund. 6. Walvis Bay. 7. Swakop River. 8. Kuiseb River. 9. Khan River. 10. Husab Mountain. 11. Langer Heinrich Mountain. 12. Marmor Pforte Ridge. 13. Schiefferberge. 14. Dune sea. 15. Coastal dune cordon. 16. Wind streaks (Satellite Remote Sensing Centre, Council for Scientific and Industrial Research, Pretoria: space image WRS no. 192-76). Reprinted with permission from Geonex Chicago Aerial Survey.*

serve known as the Namib Desert Park (figure 1), although mineral exploration and small-scale mining operations have taken place within the park.

The study area is poorly known to local townspeople, although it lies as little as 30 km east of Swakopmund. This was not the case in prerailroad times. The most important wagon route to the hinterland led from Walvis Bay across the Tumas Flats to Otjimbingwe, the Herero center on the Swakop River 100 km inland. The route thus ran northeast across the lower Tumas drainage—where it still appears to be faintly visible on space and aerial photographs—to the first watering place at a prominent granite dome next to the Tumas stream bed in the center of the study area. The route then crossed the long north slope of the Tumas valley to the Husab Gorge, at the head of which the present field camp is located. It then wound up the Swakop River valley from waterhole to waterhole. As a consequence of this route, reports and maps by several well-known nineteenth-century explorers and administrators mention the lower Tumas basin.

The British laid claim to Walvis Bay in 1876 and attached it administratively to the Cape Colony. Walvis Bay remains formally a 750 km^2 exclave of the Republic of South Africa (figure 1). Germany annexed Namibia in 1884 and constructed the artificial port of Swakopmund 40 km north of Walvis Bay. The latter was reestablished as the major port after South Africa's annexation of Namibia in 1916 (Logan 1973), and Swakopmund has become the center of tourism as Namibia's "summer capital." The town has recently grown rapidly northward with the development of the Rössing uranium mine 60 km to the northeast.

Major road and rail connections to Windhoek, the nation's capital city, skirt the study area, its surrounding rugged canyon zones, and the escarpment by crossing the Namib Desert in a northeasterly direction. Two minor dirt roads, traversing the axis of the Tumas basin, connect the coastal towns more directly with Windhoek. A maintained road connects Walvis Bay with the Namib Desert Research Station at Gobabeb 80 km to the southeast on the Kuiseb River.

Wellington (1967) and Logan (1960) have discussed the settlement history and economic geography of the Central Namib in some detail. A few Khoi (Topnaar) families live in the lower Kuiseb River canyon, and a few small white-owned farms occupy conducive points in the Swakop River valley within the park.

Chapter 2

SETTING: GEOLOGY, LANDSCAPES, CLIMATE, FLORA, AND FAUNA OF THE LOWER TUMAS BASIN

From the great length of [the Naarip Desert], and the total absence of water and pasturage, it is necessary to traverse it during [the cool of] the night. As thick fogs and mists, however, are not uncommon here, the traveler is exposed to some risk.

—C. J. Andersson, *Lake Ngami*

Geology

The ancient pre-Cenozoic and younger Cenozoic rocks of the Namib Group, divorced by a major unconformity (Namib Unconformity Surface), are described, the latter as a new group with a rapidly growing number of recognized constituent formations.

Pre-Cenozoic Rocks

The Central Namib is underlain by rocks of the Damara Province, "an integral part of the Pan African structural framework" (Tankard et al. 1982:312). The great size of the province, combined with excellent lateral and vertical exposures in the Central Namib flats and canyonlands, has drawn much attention to these rocks in the last fifteen years.

In the central and southern margin zones of the Damara Province, rocks of the Damara Supergroup comprise the following units, chronologically enumerated after Tankard et al. (1982):

1. The Nosib Group of mainly continental sediments was laid down between 900 and 1000 m.y. B.P. and 830 m.y. B.P. in structural troughs with analogs in Brazil.
2. Rocks of the Swakop Group were deposited unconformably as depository troughs widened and coalesced; thin but prominent

marbles, significant quartzites, and the very thick magnesium-rich metapelites (mica schists) of the Kuiseb Formation are included.

3. Several phases of folding with associated intrusion of granitoid rocks, manifested as the "sinuous" Damaride mountain chains, occurred between 660 and 460 m.y. ago in the Central Namib Desert. The rocks appear today as numerous outcrops of gneisses and granites of the Salem Intrusive Suite in the lower Tumas basin. Concomitantly in the upper Tumas basin, the Kuiseb Formation metapelites were tectonically deformed and thickened "by recumbent folding and imbricate thrusting" (Tankard et al. 1982: 320) on a massive scale. These make up much of the Khomas Hochland section of the Great Escarpment east of the Tumas basin and underlie much of the Central Namib north of the Kuiseb River. The 500-560 m.y.-old Donkerhoek Granites separate the Salem Intrusives from the Kuiseb metapelites and exert significant landform control in the headwaters of the Tumas basin. Prior to the emplacement of this granite, as much as 10 km of uplift was achieved (Sawyer 1981), initiating exposure of the orginally deep-lying rocks.
4. The intrusion of dolerite dikes occurred in later Karoo times (toward the end of the Mesozoic). Although individual dikes as wide as 300 m have been reported (Smith 1965), most are one tenth as wide, or less, in the lower Swakop and Tumas drainages. They form highly characteristic, low, black ridges trending north-northeast and east-northeast (Smith 1965).

Drainage patterns in the Central Namib Desert are evidently influenced by these Precambrian tectonic features. Three parallel northeast-trending grabens, located in the Central Zone of the province, have been postulated to explain the existence of the Nosib Group sediments (Tankard et al. 1982). Walvis Bay is located on the central and largest of these, and furthermore lies on the ancient north-south alignment of the northern and southern branches of the Damaran Province (Tankard et al. 1982). Basement rocks are markedly basined along the rifted continental margin west of Walvis Bay (Walvis Basin) where a pile of younger rocks several kilometers thick occurs (Dingle 1979). Bowin et al. (1981) have mapped a gravity anomaly on the coastal shelf immediately west of Walvis Bay; though small, this is the highest nearshore anomaly anywhere offshore between Cape Town and the Niger River delta.

The influence of these megastructures on modern landscapes is suggested by the fact that the angle of the great embayment of the continental margin in Namibia is centered at Walvis Bay. Furthermore, three of Namibia's largest rivers converge on the Walvis Bay sector of the coastline. Martin (1968, 1975) has mapped major Paleozoic glacial valleys in central

Namibia: these are aligned with the dominant lineaments of the Central Zone of the Damara Province and drain toward Walvis Bay, illustrating the great age and permanence of the major drainage patterns of this part of Africa. "The present-day geomorphology is largely dependant on the directions of the late Palaeozoic drainage" (Martin 1975:50).

Cenozoic Sediments: The Namib Group

The Namib Platform is usually described as decked with sheets of alluvia, fanglomerates, caliche (calcretes), gypsum crusts (gypcretes), lagoonal facies, and sand dune masses.

Such deposits are, however, sufficiently thin and discontinuous in most areas for ancient rock types to be recognizable in outcrop over most of the Namib Platform, except south of the Kuiseb River. Even where surface deposits are continuous, the flatness of the Namib Platform has disposed researchers and exploration geologists to consider them comparatively thin.

The South African Committee for Stratigraphy (1980) has recognized several discrete Tertiary Formations, including the Sossus Sand and Tsondab Sandstone Formations in the Central Namib, which overlie the ancient rocks unconformably. A spate of new units has been recognized recently, mainly centered around the lower Kuiseb and Tsondab rivers. Compared with the above-mentioned Formations which are widespread, most units are comparatively restricted areally, an observation that applies to the Tumas drainage. Possible correlations of these Tertiary deposits are discussed in the next chapter.

It is proposed here, following Hambleton-Jones (1976), that the Cenozoic deposits of the Central and southern Namib be ascribed to the Namib Group: there exist now twelve separate formations between the Tsondab and Swakop rivers, and many more beyond (see for example Ward et al. 1983). Suggested correlations of this group with the Kalahari Group are of the most general chronological nature only. The younger deposits of the Central Namib are poorly known, having been considered thin, unfossiliferous, and less economically important than the mineral-rich and structurally significant ancient rocks.

The following represents a comprehensive list of known Tertiary units in and around the lower Tumas basin (figure 3). As yet, the lithology and relationships of most of these remain obscure. It is proposed to establish the Tumas Sandstone and Leeukop Conglomerate Formations for the presently unnamed sediments central to this study, both on the strength of their newly ascertained and considerable dimensions, and owing to the fact that they are described herein according to the usual requirements (International Subcommission on Stratigraphic Classification 1976; South

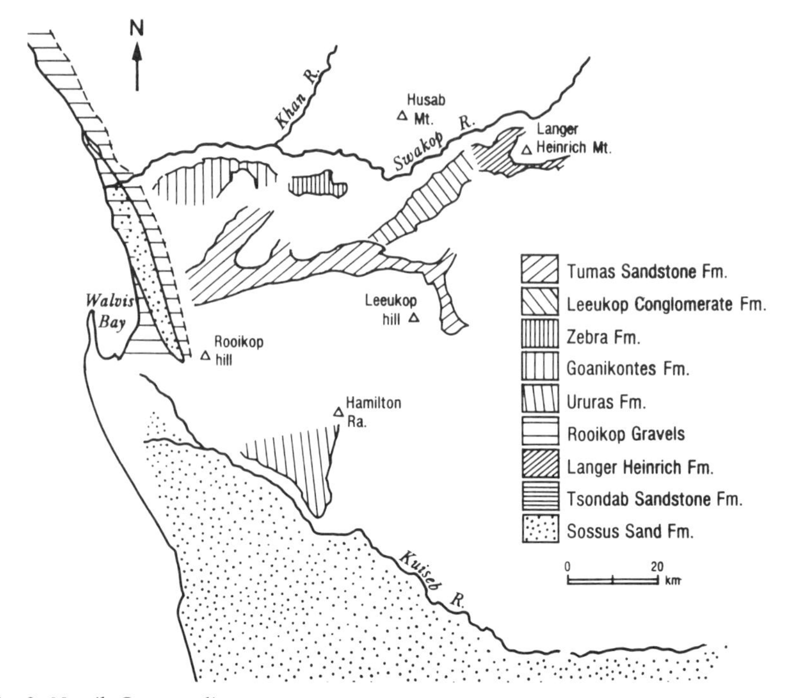

Fig. 3. Namib Group sediments.

African Committee for Stratigraphy, 1980; North American Commission on Stratigraphic Nomenclature 1983).[1]

Tumas Sandstone Formation. The Tumas Sandstone Formation is central to the present study. At least 37 km in length, it stretches from a point 20 km west of the Rabenrücken Hills due west along the 5-8 km-wide axis of the lower Tumas valley to within 15 km of the coast (figure 3). It thickens from 10-15 m in the east to 15-20 m in the west and comprises two members, a thicker sandstone with thin, overlying gypcreted gravels.

A topographically prominent deposit of uncertain stratigraphic affinity is situated 10 km south of the Tumas River course and 27 km inland. This tabular mass (850 m x 200 m) consists of red, fine to coarse sand and

---

[1] The International Subcommission on Stratigraphic Classification (1976:16-19) recommends inclusion of name, "kind and rank of unit," historical background, stratotypes and other standards of reference, description of unit type locality, regional context (especially boundaries), discussion of genesis where appropriate, geological age, references to prior literature, statement of intent, and publication in a "recognised scientific medium" of wide availability. The South African Committee for Stratigraphy (1980) has followed these precepts closely.

pebbly coarse sand with extensive gypsum veining. It may be an outlier of either the Tumas Sandstone or of the Tsondab Sandstone Formation described by Besler and Marker (1979).

Leeukop Conglomerate Formation. A thick, calcareous, sandy pebble conglomerate up to 80 m thick underlies the Tumas Formation. With no visible outcrop, data on this Formation has been derived from numerous percussion drill holes that reveal the full depth of the Leeukop as the valley fill of a buried canyon. This Formation probably underlies the 10-km-wide Gawib Flats and may underlie flanking valley side slopes of the lower Tumas valley. Its relationship to fluvial sediments at Langer Heinrich Mountain (Langer Heinrich Formation, described by Hambleton-Jones 1984) is unclear, although Hambleton-Jones (1984) has suggested that they may be lateral equivalents. Because the relevant exposures lie more than 40 km apart, these bodies are considered separate Formations at present. The westernmost 15 km of the study area and the Gawib Flats are thus underlain by large volumes of Tertiary sediment where lateral expansion of the Leeukop occurs, in contrast to the narrower central-eastern parts of the study area (figure 3).

Zebra formation. The Zebra Mountains are a short, steep-sided ridge on the south rim of the Swakop canyon due north of the study area. Immediately to their southwest lies a sediment body smaller than those mentioned above (6 km x 10 km+) (figure 3). This formation occupies the Swakop-Tumas watershed, and, as shown by drilling, varies considerably in thickness (1-45 m) as a result of highly uneven subsurface topography. These sediments are mainly sands and gypsiferous gravelly sands with gravel lenses. Near-surface sands, less than 15 m thick, are reminiscent of Member 1 sands of the Tumas Formation.

The Zebra formation is part of the extended but discontinuous train of Namib Group deposits that deck the Swakop canyon rim to west (Goanikontes formation, below), east (Gawib Flats), and north on the Welwitschia Flats.

Goanikontes formation. The Goanikontes formation is well exposed on the south flank of the Swakop canyon at Goanikontes (figures 3, 7), 10-30 km from the coast and 290-310 m above sea level. Gevers and van der Westhuyzen (1931:62) described excellent exposures of "lime-cemented feldspathic grits with a large number of conglomeratic layers and numerous isolated pebbles," which are predominantly waterworn and matrix supported (Wilkinson 1976).

The formation has attracted commercial interest. Thin (15-20 cm) lenses of pure rock salt and gypsum occur in an upper and lower set. Reuning (1925) discussed a small mined occurrence of native sulphur at the base of this formation, but exposures are few and the southerly extent of the formation in the direction of the Tumas Formation is unclear. It apparently thins to east and south where outcrops of ancient rocks occur.

Ururas formation. Widespread flat-lying sediments stretch 10 km from the lower Kuiseb River north and northeastward to the marble pediments and massif of the Hamilton Range (figure 1). These gravelly sands, with numerous discontinuous gravel beds, are heavily gypcreted and give rise to remarkably flat landscapes, best developed north of the water pumping stations of the lower Kuiseb near the place known as Ururas. Deflation hollows scoured into these deposits provide exposures showing thicknesses of up to 20 m. Cursory examination suggests that these sediments are unchannelized, entirely unfossiliferous, alluvial spreads.

Rooikop Gravels. The full extent of the coastal gravels is unknown, but they are well exposed within 10 km of the coast between Swakopmund and Walvis Bay below approximately the 100 m contour; they are discussed by Davies (1973), Wieneke and Rust (1975), Miller and Seely (1976), and the South African Committee for Stratigraphy (1980). Extending from the Ugab River mouth in the north at least as far as Rooibank on the Kuiseb River, they consist of a series of beach-related deposits as well as fluvial deposits along the Kuiseb, Swakop, and Tumas rivers, probably at least 50-100 m thick by inference from bathymetry of the mouths of the former two rivers (Rust and Wieneke 1976).

The 5-km-wide chain of dunes that stretches along the coast from Walvis Bay to Swakopmund is apparently an extension of the great dune field south of the Kuiseb River. The latter is designated as the Sossus Sand Formation (South African Committee for Stratigraphy 1980), of which the chain may be considered part. The dune chain overlies the abovementioned coastal sediments, as shown by the outcrop of bedded sands and gravels that floor deflation basins within the dunes.

Modern Landscapes

Ernst Kaiser wrote of the Plains (Flächen) Namib, apparently including the Tumas basin, as an area of mainly flat surfaces, in contrast with the Dune Namib and Trough (Wannen) Namib, in both of which eolian processes have dominated, the former characterized by great dunes and the latter by significant deflation landforms (Kaiser 1926). Gevers (1936:67) saw the Tumas basin as occupying the meeting point of "three well-marked morphological entities" of regional extent: (1) the Khomas Hochland (Highlands) to the east; (2) the "Inselberg Region of W. Damaraland" stretching south to include the mountainous northern edge of the Tumas drainage, characterized by numerous bornhardts, chains of hills and mountains, and the prominent massifs of the Brandberg (highest point in Namibia at 2,483 m altitude), Erongo Mountains, and the Great and Little

Spitzkoppe; and (3) the "Namib Desert," based on and including Kaiser's (1926) above-mentioned tripartite division. It is unclear whether Gevers (1936) regarded the relatively large Tumas basin as a flatter, transitional, southern margin of the Inselberg Region, or the extreme northern margin of the Plains Namib.

Beaudet and Michel (1978) defined a simpler division of the Central Namib, namely the northern rocky plains with the Tumas at its southern extremity, and the Namib erg. Marker (1977) distinguished between the gravel Namib north of the Kuiseb River and the Dune Namib.

The commonest descriptive terms reflect the strong climatic gradient, however. Spreitzer (1965-66) spoke of the coast or Exterior Namib where coastal effects are dominant, the Interior Namib, and the Anterior Namib against the Great Escarpment. Logan's (1960) division shows the Coastal Namib as a narrow strip along the coast, followed by the major division which he termed the "Namib Platform," transitional to the wetter interior. The terms Inner and Outer Namib, reflecting this climatic gradient, are also used, and are therefore adopted here for their usefulness and brevity.

Besler (1972) has identified three geomorphic divisions that do not coincide with those above, derived from detailed investigation of small surface features: (1) the cold foggy desert (Nebelwüste), 20-40 km wide, followed by (2) the transitional, alternating fog desert (Nebelwechsel-Wüste), 20-30 km wide, followed by (3) the desert steppe (Wüsten-steppe), approximately 80 km wide.

Logan (1960) has described in detail several repetitive landscape units within the larger entity of the Namib Platform, most of which occur in the lower Tumas basin.

The Study Area

The Tumas River rises on the Namib Platform at an altitude of 1000 m and crosses two sets of low ridges (Schiefferberg and Rabenrücken) at 550 m altitude. These ridges bisect the Tumas basin and act as the eastern boundary of the study area 80 km from the coast. Interestingly, the Tumas River from this point almost to the sea lies in a wide valley almost as far below its divides (figures 4, 5) as the Swakop River that cuts a 200-m-deep canyon into the platform (figures 6, 7). Ridge and inselberg crests rise well above platform level.

Streams on the platform generally display longitudinal profiles convex to the sky, probably as a result of discharge reduction downstream (Stengel 1964; Goudie 1972). Furthermore, gradients of most streams and their larger tributaries vary locally in a systematic way, decreasing upstream of and increasing downstream of points where the valley plain is narrowed by flanking rock outcrops. The result is a series of local convexities

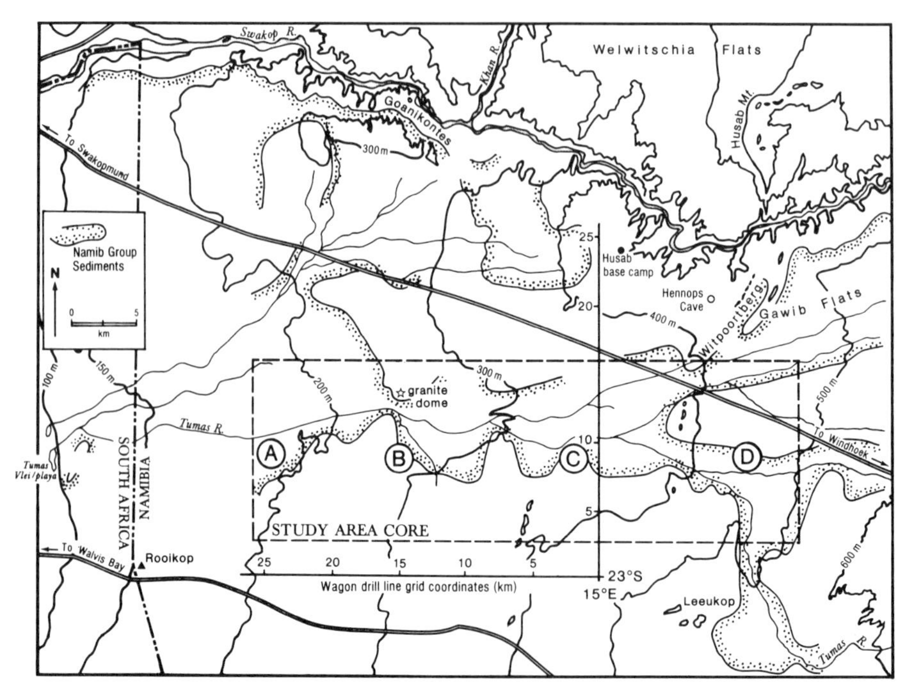

Fig. 4. Study area sectors (A-B; B-C, C-D), coordinates, Namib Group sediment outcrop, and other physical features.

superimposed on the regional gradient. Subvertical marble beds and dolerite dikes commonly approach the alluvial plain transversely from north and south, acting as more resistant lithologies that constrict valley width.

The study area is centered on a 40-km-long zone of the lower Tumas valley between 20 and 60 km from the coast (figure 1). In the eastern part of the study area the valley floor widens gradually from 1 km at the bisecting ridge, to 2.5-4.0 km downstream, to a maximum of 4.5 km in places. Long-profile gradients steepen at a dolerite dike in the vicinity of D (figure 4) from 1:130 upstream of the dike to 1:97 downstream.

The valley floor is unincised and characterized by a braiding stream habit and continuous gravel spreads.

Water table data are sparse and untrustworthy, but drilling indicated water at 6 and 10 m in the vicinity of the dolerite constriction, declining westward to 14 and 25 m at 6 and 10 km respectively downstream from the dike.

Long profiles flatten rapidly to 1:200 in the center of the study area around the small dune field. Further west, slopes of the highest of three surfaces developed on the alluvial plain are even gentler (1:400). By contrast with the eastern sector, the Tumas River is incised as much as 20 m below the upper surface and is characterized by two major, slightly sinuous arms. In contrast to the eastern end of the study area, these arms are never more than 150 m wide and are walled, vertically on outer bends, by more coherent gypcreted sediments. The north arm gathers a tributary draining the Gawib Flats.

Valley sides in the central and western parts of the study area are incised very little with drainage lines usually less than 2 m below surrounding surfaces.

Water tables in the western sector rise slowly from a 20 m depth, reaching close to the surface in the six westernmost kilometers of sector B-C (figure 4). Standing water has not been encountered, however. Water tables decline again progressively with distance west from B for at least the next 15 km, reaching known depths of 55 m.

The valley sides in the central sector comprise pediments of very low declivity (<2°) and very little relief. Pediments are cut in granites, gneisses, and quartzites, the less resistant rocks in the area. In places, broad sandy

Fig. 5. Tumas Flats between Rabenrücken (background) and Swakop River canyonland margin (foreground). Black dolerite ridge foreground. Tumas River and marble inselberg far middleground. View southeast.

Fig. 6. Swakop River canyonlands ("grammadullas") exposing Damara Supergroup metamorphic rocks. Tumas Flats/Namib Plain on canyon rim skyline. Swakop River middleground. View southeast.

washes cover pediment surfaces thinly, or in the case of larger washes, thickly, since the latter washes are associated with buried drainage lines of a hillier, prior topography.

Many pediments, however, are entirely rock-cut and display a discontinuous veneer of small exfoliation scales and grus particles. Pediment surfaces are scored by continuous, narrow, and abrupt rills 10-20 cm deep, which can be damaging to vehicles. More resistant dolerite dikes produce local relief of less than 3 m, such is the degree of planing that has been achieved.

The eastern part is somewhat hillier and washes are incised 1-5 m and in a few locales, display tafoni and solutionally deepened hollows up to 10 m across.

Three granitic domes, one in the Tumas valley plain (B, figures 4, 22), and the other two on the south flank, reach heights of approximately 50 m. The existence of these few domes does not materially alter the generalization that plutonic rocks on the platform are less resistant.

Nosib quartzites are somewhat more resistant and form low ridges up to 100 m high on the south side of the valley.

Steeply dipping (40-80°) white marble beds are the most resistant lithology and project above the platform as the most prominent inselbergs, two in the eastern sector rising 100 m (Leeukop hill) and 170 m above platform pediments. More commonly, thin bands of marble give rise to long, sinuous, sharp-crested ridges of lesser height. Slopes on the marble are commonly 10-35°, although the steepest slopes on the granite domes attain 60°.

Marble ridges crop out particularly on the south flank of the Tumas valley, cutting across pediment surfaces. By contrast, the longer north flank between the Tumas and Swakop rivers is nearly devoid of marble beds, so that the pedimented landscape is almost flat, despite being laced by a network of near-vertical dolerite dikes. Slopes are less than 1°.

Topographic maps suggest the existence of a few small alluvial fans as opposed to pediments, on the west flank of the Schiefferberg. The lower Tumas basin, however, is devoid of these classic arid-zone features, as Logan (1960) has noted. Even footslope accumulations of debris are uncommon, so that the study area has a rocky, angular aspect of low-angle ped-

Fig. 7. Larger valley in left bank canyonlands of Swakop River: canyon rim (Namib Platform) developed on thin sabkha deposits of the Goanikontes formation (horizon, left); Rössing Mountain background. View northwest.

iments or depositional surfaces, punctuated by abruptly steeper, gravity-controlled slopes. Despite hyperaridity, slope profiles are locally remarkably convexo-concave (nonangular) in morphology where developed on young, gypcreted Namib Group sedimentaries (in the western sector of the study area).

Pediment slopes shelve beneath thickening sediments of major drainages such as the Tumas River and its tributaries. Gypsum crusts in various forms blanket all lower-lying surfaces and take on geomorphic importance. As a rule, crusts occur on all slopes that are mantled with less coherent materials such as regolith, lower pediment sediments, and alluvia of the Namib Group sedimentaries, provided these materials are undisturbed. A near-surface, meter-thick accumulation of amorphous gypsum, underlying tens of hectares of the Tumas alluvial plain near the dune field, has been mined in a small way.

Patterned ground and associated vegetation polygons are common on duricrusts near the Kuiseb River and in the interdunal areas of the dune sea (Goudie 1972; Watson 1980), but are not widely developed on the Tumas floodplain.

Near the Swakop canyon rim, a low marble outcrop hosts a small, near-surface cave 12 m high and 12 m long of apparantly precanyon age. It is floored by young, water-laid sediments rich in microfaunal remains (Wilkinson 1979).

Climate

The climate of Namibia is dominated by two quasi-stationary anticyclones, one located over the African subcontinent between approximately latitudes 24° and 27°S, and the other over the southern Atlantic Ocean centered between latitudes 27° and 30°S (Jackson and Tyson 1971). Winds from the SSW blow throughout the year (Royal Navy and South African Air Force 1944). These winds drive the north-flowing Benguela Current and generate maximum upwelling of cold bottom water adjacent to the coast (Bang 1973; Andrews and Hutchings 1980; Brundrit 1981). Winds are highly variable inland, however, northern quadrants being dominant (Breed et al. 1979). Wind regimes compare with most other deserts and belong to the intermediate energy group. The Kalahari regimes on the plateau display low energy wind environments (Fryberger 1979).

The Namib thus occupies a zone of interaction between the major anticyclonic circulation systems. The southerly quadrant winds have generated a narrow (40-50 km) strip of active barchanoid dunes along the western edge of the Dune Namib. In contrast, eolian sand stringers and streaks in the study area 40-100 km from the coast are generated by a strong, hot, and dry wind locally termed the "Berg Wind," which blows from the east and northeast. These winds appear to rise when the continental anticyclone weakens and a low pressure system approaches the subcontinent from the southwest (Jackson and Tyson 1971). These winds occasionally give rise to dust storms in the Namib.

Along Africa's desertic southwest coast temperatures are strongly controlled by proximity to the Benguela Current: annual means are many degrees lower than zonal equivalents and increase slowly northward from 14°C at Port Nolloth (latitude 35°S), 1,100 km south of Walvis Bay, to 22°C at Luanda (latitude 9°S), 2,300 km north of Walvis Bay (Meigs 1966).

Temperatures are strongly oceanic with very low seasonal and daily ranges (5° and 8°C respectively) (Schulze and McGee 1978). Freezing is thus rarely experienced. Logan (1960) cited extremes of -3.9°C and 40°C for Walvis Bay. Overtly marine effects are felt only in a coastal tract 40-50 km wide (Meigs 1966). It is generally accepted that mean daily temperature ranges increase inland from the coast: 8.3°C at Walvis Bay to 18.2°C at Gobabeb, 56 km from the coast (Besler 1972) and 15.5°C at Goodhouse in the southern Namib 140 km from the coast (Meigs 1966). But Besler (1972) has shown that in the Central Namib temperature ranges are greatest at Gobabeb midway between the coast and the escarpment. Besler (1972) explained this phenomenon in terms of insolation attenuation by cloud, (1) at the coast as a result of fog banks, and (2) over the escarpment to the east as the result of increasing frequency of orographically induced cloud.

Differences in temperature can be noted within short distances of the coast. Walvis Bay is distinctly more continental than Swakopmund because it is situated on the inland side of the bay, fully 10 km from the shoreline of the great Pelican Point sandspit. Temperatures also rise gradually with distance seaward: 280 km from the coast, averages are several degrees higher (Currie 1953). The coastline thus acts as the axis of low temperature in the region.

As phenomena of the winter months in particular, Berg Winds modify this continental effect in a strange way: absolute temperature maximums occur in winter for coastal weather stations (Swakopmund, Walvis Bay, and Rooibank), whereas summer maximums occur for those stations inland of Rooibank where land-generated effects outweigh those of the Berg Wind. The Berg Wind blows most often on the west coast of southern Africa, averaging fifty times per year (Jackson and Tyson 1971).

The precipitation gradient resembles that of temperature in some ways. The coast is probably also the center of aridity (Lydolph 1973), imposing a north-south trend on isohyets along hundreds of kilometers of the subcontinental coast. Logan (1960) has illustrated the east-west gradient along the north side of the Tumas basin. The 42-year record at Swakopmund gives an annual mean of 13 mm—or 16 mm if the catastrophically wet year of 1934, when 156 mm fell, is included. The shorter record at Donkerhoek 135 km inland gives an annual average of 166 mm. Twenty years of recording indicate precipitation maximums in summer at Walvis Bay and an annual total of 23 mm (Meigs 1966). Rainfall is highly irregular, although sufficient pattern is recognizable to allow Logan (1960) the observation that rains fall mainly in late summer.

A two-year record at Walvis Bay shows a total precipitation of 31.9 mm, of which only 12.4 mm was rainfall; of 80 mm for Swakopmund, only 44 mm was rainfall (Besler 1972). These figures illustrate the importance of including fog-derived precipitation in gauging total precipitation. Boss (cited in Eriksson 1958) measured yearly fog precipitation amounts of 35.6 and 40.6 mm on horizontal plates at ground level. Meigs (1966) commented that vertically oriented surfaces would produce even more condensation. Using vertical-sided traps, Lancaster et al. (1984) documented the fact that fog can provide substantially more moisture than rainfall in the Tumas basin (183 versus 21 mm at Vogelfederberg station, and 17 versus 80 mm at Rooibank station). The effects of fog-moisture supply on soil development are examined in chapter 7. Plants undoubtedly gain moisture from the fog, as do dune-dwelling tenebrionid beetles (Seely and Hamilton 1976).

Fog frequency drops off inland, but present data is inconsistent. Goudie (1972) cited an annual average of 102 days in a four-year period at Gobabeb, 56 km from the ocean. Logan (1960), however, mentioned that fog occurs as many as 300 nights per year at Rössing 40 km inland, and Meigs (1966) in one year measured dew or fog on 285 days at the coast north of Lüderitz. An explanation for these discrepancies may lie in the topographic rise of the Namib Platform. The above-mentioned stations of Vogelfederberg and Rooibank have far higher annual fog precipitation totals than stations either on the coast or farther inland. This suggests that coastal stratus impinges on the comparatively steep Namib Platform preferentially at the altitude of these stations.

Another explanation has been invoked by Logan (1960) who identified two modes of fog formation, one advectional at the coast and the other radiational in a belt approximately 30 km inland. Nevertheless, Walter (1936) saw an inland limit of "fog-fed" vegetation 55-60 km from the coast.

Topography determines the distribution of fog in part, with valleys fogbound when higher ground is clear. Fog banks remain significantly longer and penetrate farther inland up the Tumas valley than along the flanks of the valley. Probably more effective than fog duration, however, is positive topographic relief in generating fog precipitation. Even small ridges 1-2 m high are markedly wetter after fog than surrounding flat surfaces, and vegetation density is significantly higher.

Humidity patterns follow fog frequency patterns and are very high at the coast in both seasons, minimums at Swakopmund averaging 90% (Logan 1960) in what Logan (1960:60) has called this "cool fog-shrouded, dripping desert."

Besler (1972) and Barnard (1965) have given similar Köppen classifications of Central Namib climates. Cooler, BW kln coastal types (average annual temperature <18°C) give way to warmer, summer rainfall, BW hw types in the Inner Namib. To the latter, Barnard (1965) has added the designation for spring maximum temperature (BW ht's) and Besler (1972) interposed a transitional zone between the two, stretching from as little as 20 km to as much as 80 km from the coast, which is both warmer than the coast but often foggy (BW hn). Meigs (1966) designated the Central Namib as an extremely arid desert with equable temperatures (Ea 22), grading into an arid, summer precipitation desert with an increased temperature range (Ab 23).

Flora and Fauna

Schulze and McGee (1978) applied the system of Holdridge Life Zones to southern Africa, assigning the Central Namib (up to 80-100 km inland) to the Desert Life Zone, subdivided west to east into (1) warm temperate parched, (2) superarid subtropical, and (3) superarid tropical. The Desert Zone is followed inland by the Thorn Woodland Life Zone. Water surplus in the Desert Life Zone is below 100 mm and annual water deficiency is 600-799 mm (Schulze and McGee 1978). Because of fog precipitation, this deficiency is less than that of the "edaphic" desert to the east—as Logan (1960) described the sandy Kalahari desert surface—and less, too, than those of the valleys of the Zambezi and Limpopo rivers, despite the latters' high absolute annual precipitation.

Werger (1978) has argued for a Karoo-Namib Region in his "phytochorological zonation" of southern Africa, comprised mainly of an "open dwarf shrub formation." Woodlands and savannas of the Sudano-Zambezian Region are dominant inland of a diffuse transition zone on the Hochland to the east (Werger 1978).

Werger (1978) further subdivided the region in the Central Namib: the drier, very open Namib Domain vegetation gives way to the grassier and more wooded Namaland Domain of the escarpment zone. Nel and Opperman (1985) provided an eight-part division of the Central Namib Plains north of the Kuiseb River. The lower Tumas basin lies mainly in their *Salsola* sp. desertic dwarf shrubland of the gravel plains. Robinson (1977) recognized four predominant habitats in a more detailed study of the Central Namib: (1) salt marshes, (2) pans, (3) sand dunes, and (4) the "plains-washes-rock" congerie of surfaces.

Low ridges in the Namib are inherently better watered and give rise to taller shrubs such as Commiphora and Euphorbia (Moisel 1975). The well-known dioecious gymnosperm *Welwitschia mirabilis* is a low plant with two long (2-3 m), broad leaves that grow from a mainly subterranean woody trunk up to 1 m in length (Logan 1972). In the Central Namib these plants occupy the beds of washes in the drier western areas.

More than 80 km inland, low isolated acacia trees begin to appear in washes, increasing to thin, discontinuous lines of trees further inland. Sparse tree vegetation can be seen in the steep side valleys of the Kuiseb and Swakop river canyons only 50 km from the coast.

By contrast, in the canyon bottoms where ground water is plentiful, thick stands of *Acacia erioloba* and *A. albida* make up a community with *Tamarix usneoides* and many other shrubby species. These stands reach almost to the sea, dying out probably as a result of increasing water salinity.

In terms of zoogeography, the Namib Desert belongs to the Karoo-Kalaharian Subregion of the Ethiopian Region in Franz's (1970) interpretation. The subregion occupies southern Africa with the exception of the moist east and south coasts. The number of species is very low in the western Tumas basin. The commonest mammal species are the oryx, jackal, and hare, usually seen as individuals, and small groups of springbuck and zebra. Larger bird species are owl, vulture, and kori bustard, the last being one of the heaviest flying birds in the world. Horned adders seem to be the commonest snake in the Tumas valley.

Inland of the Rabenrücken-Schiefferberg hills where grass growth is far more plentiful, the number of larger animals increases dramatically so that small herds of the above-mentioned ungulates are encountered. The horn of a rhinoceros was found on the Gawib Flats (Sauer 1972), and elephants are known to have wandered the length of the Swakop River as far as the coast. Nowadays desert elephants wander the Kaokoveld plains farther north (Ward et al. 1983).

Chapter 3

CENOZOIC GEOMORPHOLOGICAL AND GEOLOGICAL EVOLUTION OF THE CENTRAL NAMIB DESERT

Africa . . . that most repulsive . . . and least accessible quarter of the globe.

—C. J. Andersson, *The Okavango River*

Second, you neglect and belittle the desert.
The desert is not remote in southern tropics.

—T. S. Eliot, "The Rock"

This chapter provides a regional context for study-area landscapes and the Namib Group deposits in terms of both external stratigraphic connections and possible chronologies of Tertiary events. I shall discuss the morphology of the major unconformity separating younger Namib Group sediments from the ancient basement rocks, as well as related questions of drainage pattern evolution in the Central Namib. I shall also consider Namib Group sediments in the Tumas basin in relation to better-documented successions further south; and, finally, I shall present two possible chronologies.

A full picture of Tertiary events in the study area ought also to account for certain phenomena idiosyncratic to the area, in particular, (1) the buried canyon beneath the present lower course of the Tumas River; (2) the underfit nature of the lower Tumas River, which occupies a basin so large (23-27 km wide and 75-150 m deep) as to appear deceptively flat; (3) the present course of the lower Swakop River valley, which cuts across several major geological structures and the associated topographic barriers, bypassing unbarricaded routes to the sea; (4) the presence of undoubted lagoonal facies at 300 m and more above sea level; (5) the paradox of apparently different-aged sediments at similar morphostratigraphic positions on the south rim of the Swakop River canyon; and (6) the supply of thick masses of sediment along the entire coastal tract between the Kuiseb and Swakop river mouths, especially in the lower 20-25 km of the Tumas basin.

Namib Unconformity Surface and Drainage Evolution

Ollier (1977:207) has termed the Namib Unconformity Surface (NUS) "a fundamental datum . . . separating the metamorphic bedrock from all younger deposits." This statement suggests that the NUS has chronostratigraphic importance. In fact it has suffered major modification since the first emplacement of postmetamorphic deposits, probably in the very early Tertiary, and continues to this day to undergo change where it is exposed. That the NUS is a polygenetic feature is evident from the fact that it has been incised and rebeveled on two occasions at least, and is today undergoing pedimentation where it is exposed in the study area and elsewhere—for example, in the inner Central Namib (Rust 1970) and in the Ugab River valley, where two phases of post–"Pedestal Surface" modification have been documented (Mabbutt 1952). Sea-level fluctuations and epeirogenic uplift have generated a wide coastal tract where marine, coastal, and terrestrial environments have all held sway at different times.

At its most ancient, the NUS is often equivalent in the southern Namib to the "end Cretaceous land-surface," where it is dated by the overlying Paleocene Chalcedon-Tafelberg Silcrete Formation. In southern Angola, the late Cretaceous Giraul Conglomerates (Ward et al. 1983) overlie the surface. Researchers have reasonably assumed that the planar NUS in the southern Namib "may be tentatively correlated with the well-planed bedrock platform in the Central Namib" (Ward et al. 1983:177). Although the NUS in the Central Namib is shown here to be anything but well-planed, except where it is exposed, there is little doubt that the statement is true.

Stocken (1975) suggested the possibility that the Swakop River may have flowed in a southwesterly direction from Langer Heinrich Mountain, debouching in the vicinity of Walvis Bay (figure 1) 25 km south of its present mouth. By this alignment, such a proto-Swakop River would have crossed the Gawib Flats (now occupied by a nonfunctional northern tributary of the Tumas) and then flowed down what is today the western part of the lower Tumas basin. The upper Tumas River would have acted as a south-bank tributary entering the proto-Swakop River in the center of the study area at approximately the 320 m contour. Reorientation of the lower courses of west coast rivers is reported in the cases of the Orange and Olifants rivers (Hendey 1983).

Stocken's (1975) hypothesis explains several facts concerning the NUS, in particular the existence of a valley 10-25 km wide, cut about 100 m below the surrounding levels of the Namib Platform. This shallow trough is presently occupied by 50-120 m of sediment and has given rise to a topographically flat tract of country all the way from Langer Heinrich Mountain to Walvis Bay. Second, the hypothesis explains the presence of a particularly wide tributary valley entering the lower Tumas on its north bank near the Tumas Vlei. As a linear extension of the present Khan River canyon, the tributary appears to serve as a lower proto-Khan River (figure 4). Third, the present course describes a distinct change in direction at the mountain (from southwesterly to westerly), suggesting an elbow of capture (figure 1).

The hypothesis does not explain why the Swakop River should have switched to this more northerly position directly transverse to several ranges of hills of resistant lithologies. But it does account for the underfit nature of the Tumas River and major tributary in their lower courses. It also explains (1) the 80 m+ thickness of fluvial deposits that crop out along the south side of the Swakop River canyon between Langer Heinrich Mountain and the Witpoortberg ridge (figure 1), (2) the great mass of gritty deposits in the coastal tract, and by the same token (3) the existence of what appears to be one of the largest buried canyons in or around the study area. The hypothesis holds even if crustal downbowing at major depositories such as river mouths (Hallam 1964) is taken into account.

In an erosional landscape where the local maximum topographic variation on the remarkably planar Namib Platform is less than 5 m, the buried NUS topography cut in the ancient rocks is notably variable: e.g., it is intercepted at 45 m depth beneath the Zebra formation, at 80 m depth north of the Chuos Mountains (on the farm Valencia), at 100 m in the study area, at 115 m near Garub in the southern Namib Desert (see Range, in Gevers 1936), and at 100-200 m in the Ubib drainage east of the study area (Hüser 1976). These suggest that the NUS (40-80 km inland at least) has a relatively variable topography. If the present exposed pedimented topography is taken into account, the NUS is yet more varied altitudinally in the study area (up to 270 m). Extensive drilling of the Tumas, Leeukop, and Zebra sedimentaries shows that the NUS, where buried, is better characterized as rugged, since it displays very abrupt slope changes (figure 8).

The ages of the buried sections of the canyon and the overlying sediments are best understood in light of the improved documentation of coastal uplift and major eustatic episodes of the Tertiary and late Cretaceous (see review in Dingle et al. 1983). The now-buried Tumas canyon is most plausibly related to either or both regressions of the earlier history of the subcontinent, namely the Maastrichtian and Oligocene, when base levels lowered for millions of years probably allowed incision in all exoreic rivers.

Considerations of Regional Stratigraphy

Not only is the sequence of Tertiary events equivocal in the Tumas basin, but external correlations remain speculative. Nevertheless, the Tumas succession shows apparent parallels with the major features of the Namib Tertiary succession, a succession synthesized briefly by Martin (1950). This scheme, with its proposed stratigraphic correlates spanning the length of the Namib Desert as well as certain Kalahari units, has been followed

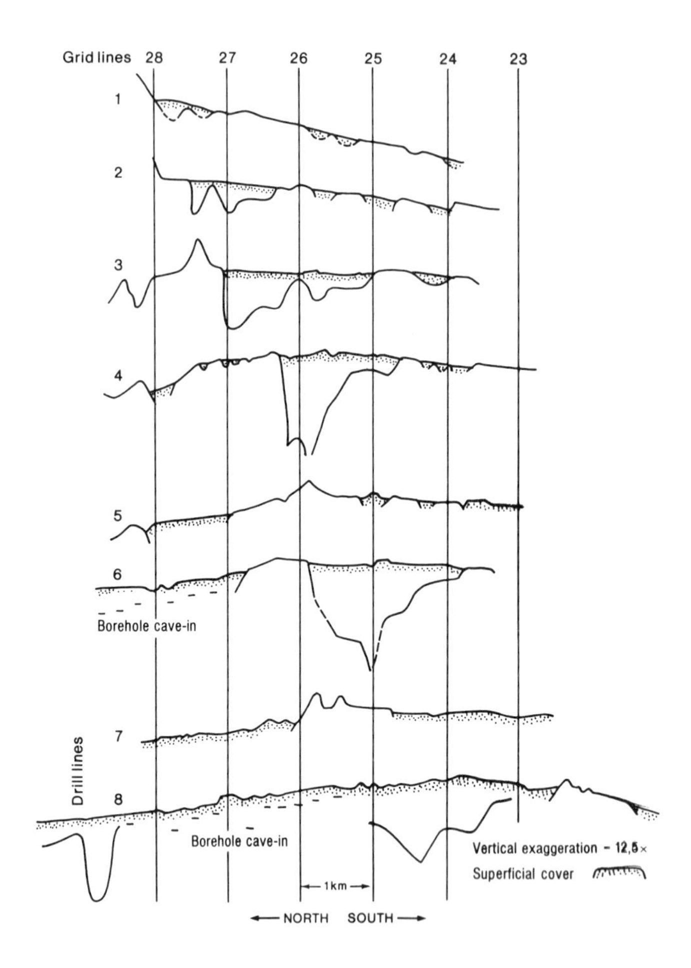

Fig. 8. Namib Unconformity Surface (NUS), planed where exposed at surface, relatively rugged where buried. Vertical exaggeration 25x.

with little modification by Ward et al. (1983). The major features of this regional sequence are a basal breccia or conglomerate overlying unconformably the ancient pre-Cenozoic rocks (Beetz 1926; Martin 1950; Ollier 1977, 1978) of probable pre-Eocene age (Ward et al. 1983). The basal unit is in turn overlain by well-developed sediments of continental origin, typified by the early to mid-Tertiary Tsondab Sandstone Formation (Ward et al. 1983). Gravels of Middle to Upper Tertiary age cap the succession in many places. Calcium-rich duricrusts have indurated large areas, acting as something of a geomorphic marker lithology.

TABLE 1.
Namib group sediments in and around the Tumas Basin.

	Central Tumas Basin	Tsondab drainage and other areas (after Ward et al. 1983, Ollier 1977)
Holocene	small dune field	dune evolution
Pleistocene	gypsification (S2-S4 crusts); stream bank colluvia	Hamilton Vlei conglomerate; Oswater conglomerate; Homeb silts; dune evolution; calcification (tufa, pedogenic- and groundwater-emplaced)
Plio-Pleistocene		coastal dune cordon and Sossus Fm. (dune sea sands)
Miocene	first gypsum crust (S1); uranium emplacement in Tumas Fm.	Kamberg Calcrete
Miocene	Tumas Sandstone Fm. Mb. 2 (gravel sheet)	"high terrace conglomerates" (Tsondab, Kuiseb, Khan, and Swakop drainages)
Miocene	Tumas Sandstone Fm. Mb. 1 (red sandstone)	upper levels of Tsondab Fm.; evolution of Tsondab Planation Surface
Eocene-Miocene		Tsondab Sandstone Fm.
Oligocene	Reorientation of proto-Swakop River	
Eocene	Goanikontes fm. (sabkha sediments)	
Eocene	Leeukop Conglomerate Fm.	Basal Breccia/Conglomerate
	Namib Unconformity Surface (NUS) cut as rugged valleys	NUS

The succession of Namib Group sediments in the lower Tumas basin suggests the following prima facie parallels with this generalized sequence (table 1). The lowest sediments altitudinally are the thick conglomerates buried in the canyon. These may correlate with the above-mentioned "basal breccia" and subsequent transgression-related sediments. Member 1 sands of the Tumas Formation suggest parallels with the widespread mid-Tertiary Tsondab Sandstone Formation south of the Kuiseb River. The Member 2 conglomerate may correlate with the conglomerates exposed along many canyon summits. Duricrusting of exposed younger sediments in the Tumas basin by carbonate and sulfate compounds suggests further parallels.

Several researchers have promoted aspects of the sequence outlined above (Beetz 1926; Gevers 1936; Martin 1950; Ward et al. 1983). Although the Tumas succession appears alluringly similar, the fact of the remarkable change in lithology and topography on either side of the Kuiseb River, and the widely acknowledged difficulty of drawing passable stratigraphic parallels between younger sequences, even in neighboring valleys, require closer examination of the chronology and external correlations.

Leeukop Conglomerate Formation

The conglomerates that occupy the Tumas canyon may be the equivalents of the above-mentioned basal conglomerate, lying as they do immediately above the NUS. An Eocene age is proposed for the conglomerate, which gibes with alluviation associated with the Palaeogene marine transgression complex. An Upper Miocene date—coincident with the major trangression of the time—might apply if the canyon were of Oligocene age.

Goanikontes formation

Despite the vagaries of relative sea-level fluctuations, the generalization holds good that altitude correlates crudely with deposit age. Marine sediments of the lower Buntfeldshuh Formation in the southern Namib and at Needs Camp on the southern Cape coast are probably of Eocene age (Siesser and Salmon 1979; Siesser and Dingle 1981). Since these deposits lie at altitudes of 120-170 m and ca. 360 m respectively, it seems likely that the Goanikontes lagoonal sediments—at 290 m+ above sea level—may well relate to the period of Eocene sedimentation along the continental margin.

Dingle et al. (1983) show that the highest marine sediments along the southern and western coasts of southern Africa correlate best with an earlier transgression in the latest Cretaceous. It is not impossible that the Goanikontes formation is associated with this ancient trangression. If it is, then it rivals the Chalcedon-Tafelberg Silcrete Formation at Pomona in the southern Namib Desert as one of the oldest post-NUS deposits on Africa's southwestern margin.

Miocene and Pliocene marine terraces of "remarkable continuity" (Dingle et al. 1983:282) crop out along the west coast between Port Nolloth and Varswater at altitudes of 20-90 m, suggesting that Neogene benches lie below +100 m. Stocken and Campbell (1982) have reassigned the diamondiferous D-suite beaches (22-35 m above sea level—Dingle et al. 1983) of the Orange River mouth to the Miocene, based on constituent thermophilous fauna. Attempting extrapolations northward, Kent (in South African Committee for Stratigraphy 1980:609) suggested that the marine Rooikop gravels near Walvis Bay, at ca. 25-45 m, "probably correlate with the 30-m-gravels at Oranjemund." Marine limestones assigned Miocene-Pliocene ages achieve altitudes of +300 m in the eastern Cape (Siesser 1972), but in the light of the opinions and evidence mentioned here, it seems unlikely that the Goanikontes formation at 300 m is a correlate. It seems rather to relate to transgressions that predate the late Miocene.

Oligocene–Early Miocene Incision

Geomorphic mapping of the Namib Platform in and around the lower Tumas basin shows clear evidence of planar rock-cut surfaces, features that Gevers (1936), Spreitzer (1965-66), and Goudie (1972) have noted. Two major surfaces are evident in the lower Tumas basin, the upper represented by the canyon rim of the Swakop and Khan rivers, and the lower by side slopes adjacent to the Tumas valleyway. Gevers (1936:78) has commented on the existence in the southern Namib of "a well marked peneplain into which the rivers of lower Miocene age incised their beds." Similarly, it is possible that the major marine regression that culminated in early Miocene time (Dingle et al. 1983) may have led to entrenchment by the proto-Swakop, especially in the lower Tumas valley, to the order of at least 200 m. The Goanikontes formation, plausibly Eocene in age, now lies 50-100 m above the altitude of the present Tumas River bed, above a well-marked topographic break.

The present landscape in its broader perspective is thus a mosaic of both sedimentary infilling of prior rugged topography and subsequent beveling (to levels imposed by the trunk stream) of both the younger clastics and the pre-NUS metamorphic rocks. Gevers's (1936:73) description of the evolution of the Namib Platform is therefore partial and applies only to parts of the study area generally nearer the coast: "The products of rock disintegration gradually accumulate, burying the area in its own debris, until only elevations composed of particularly resisting [*sic*] rock-types still project above the general plain."

Detailed geomorphic and geological mapping in the central and western sectors of the study area have shown conclusively that the Namib Platform is cut dominantly in pre-Tertiary rocks; larger bodies of Tertiary sediments are generally confined to valley axes of the earlier rugged landscape and become widespread only along the 5-km-wide Tumas and 10-km-wide Gawib alluvial plains. Only in the western sector and coastal tract do the younger sediments mask the older rocks almost completely (e.g., Geological Survey of the Republic of South Africa and South West Africa/Namibia 1980).

Thus the broad valley of the lower Tumas basin is less a product of regional aggradation than entrenchment and subsequent valley widening, cut across the widespread ancient rocks and local valley-fill alluvia. Even in the coastal tract where the ancient rocks remain buried, erosional forms are the norm for mesoscale features.

The noted flatness of the Namib Platform in the study area is thus a product of subaerial erosion comprising rock-cut pediments planed to a remarkable degree of perfection. The present evenness of the plain is not predominantly the result of burial beneath sheets of autochthonous debris, as

Gevers (1936) suggested, except in the lower courses of the few largest rivers of the Namib Desert. Published geological maps covering the Namib Desert show much detail of the ancient rocks (Geological Survey 1980) suggesting that these generalizations indeed hold true for the many small drainage basins that most typify Namib landscapes.

Tumas Sandstone Formation

On the grounds of (1) significantly lower consolidation, (2) higher stratigraphic position, and (3) a prominent buried shoulder at approximately the lower boundary of the Sandstone, it is proposed that this Formation is significantly younger than the underlying Leeukop Formation. The Tsondab Formation is poorly dated (Ward et al. 1983), so that the Tumas Formation could well be coeval with it (table 1), especially with its latest phases. The fact that the Member 1 sandstone apparently does not extend from the Tumas proper across the Gawib Flats to the outcrops on the Swakop canyon suggests that either the proto-Swakop River had by the time of Member 1 (?Pliocene) diverted to its present course, or that the supply of sand originated nearby in the Tumas basin. The fact that the sands of the Tumas Formation are exclusively fluviatile, when it has been generally accepted that Tsondab Sandstone is dominantly eolian, does not necessarily exclude correlation of the two, as is argued in a later chapter.

In a similar way, the upper, gravel member of the Tumas Formation (Member 2) matches the final unit of Martin's (1950) regional sequence. Morphostratigraphically at least, the Tumas gravels appear to be the equivalent of the canyon rim conglomerates (of the present Swakop canyon, at least at Langer Heinrich Mountain) that Martin (1950) tentatively assigned to the Pliocene and more recently as possible equivalents of the fossil-rich mid-Miocene Arries Drift Gravels (Martin, in Ward et al. 1983). Gevers (1936), on the other hand, has suggested equivalences of the rim gravels with the lower Miocene, fluviatile sediments of the southern Namib.

Possible Chronologies

Two chronologies are plausible for the events described above, since stratigraphic continuity is insufficient to erect a complete relative chronology. Suggested dates rest on the assumption that the Goanikontes formation is most likely Eocene in age.

The first accords better with regional sequences and inferred paleoclimates. (1) A rugged Namib Unconformity Surface (NUS) is fluvially cut, including the particularly deep Tumas canyon. (2) Widespread sedimentation by the proto-Swakop, Khan, and Kuiseb rivers buries most of the ancient rocks 30 to 70 km inland in the lower Tumas area; this includes deposition of the Leeukop Formation and the Tumas canyon fill. Stage (2) ends with deposition of the coastal (?Eocene) Goanikontes sabkha facies, which presently lie 300 m above sea level and as little 20 kms from the coast. (3) The proto-Swakop incises a broad valley into these deposits, along the axis of the lower Tumas, in response to the worldwide late Oligocene–early Miocene marine regression. Incision occurs to approximately the level of the present Leeukop–Tumas Formation contact ca. 100 m below the Goanikontes deposits, and their correlatives, on the Tumas basin divide. (4) Reorientation of the Swakop River occurs and alluviation by the Tumas River emplaces the Tumas Sandstone Formation. Planation of large areas of pre-Tertiary rocks in the lower Tumas basin continues, producing very even surfaces. The events discussed in the present study have taken place against this background of landforms and lithologies.

The alternative, fivefold sequence is the less likely. (1) The NUS is developed as a flat to undulating surface in conformity with the "impressively level" Cretaceous surface in the Southern Namib (Ward et al. 1983:177) and the planed surface visible along the south side of the Kuiseb River canyon. (2) Upon this is laid a series of deposits—the first of the Namib Group—culminating in Eocene times with the Goanikontes coastal sediments. (3) Thereupon intense dissection by the proto-Swakop River in the lower Tumas basin scours out the Tumas canyon, incising first the overlying deposits and then the basement rocks, perhaps to a depth of 50-200 m (late Oligocene to early Miocene?). Tributary incisions related to both the proto-Swakop and proto-Khan rivers may have reduced divides between the two watersheds, approximately along the line of the present lower course of the Swakop River. (4) The lower Tumas valley is alluviated approximately to present levels by sediments that have been designated the Tumas and Leeukop Formations, possibly coincident with the late Miocene marine transgression. (5) This level may have been sufficiently high in the vicinity of the Husab-Witpoortberg area to allow the Swakop River to reorient its lower course across the Swakop-Khan divide into its present shorter, west-flowing route. Planation of ancient rocks occurs with respect to levels of alluviation established in event (4). Events external to the Tumas valley, such as the incision of the Swakop, Khan, and Kuiseb canyons, must postdate, first, emplacement of the Goanikontes deposits that they truncate, and second, reorientation of the Swakop and Khan rivers. In the Tumas basin, probably endoreic once severed from the Swakop River at reorientation, there appears to have been no response to the tectonic events that induced canyon incision in the neighboring allochthonous basins.

Mabbutt (1952) suggested a sequence of events for the Central Namib in which a shallow valley depression in the early ("African") erosion surface subsequently experienced a phase of deep incision. This sequence seems less plausible in the case of the Tumas River since there remains no evidence of the earlier valley-bottom terrace deposits.

One of the six problems outlined above remains unaddressed, namely the paradox of different-aged sediments (?Eocene Goanikontes Formation versus the Miocene or even Pliocene age of the Gawib canyon-rim deposits) occupying similar morphostratigraphic positions along the Swakop River. The elision of surfaces in flat landscapes and across erodible alluvia probably provides the explanation, giving the impression of similar ages for sediments that outcrop along canyon rims.

Erosional features that may recognizably bridge the Kuiseb canyon zone and allow the chronologies of the Kuiseb and Dune Namib to be connected to those of the study area are (1) the NUS, (2) widened valleys let into the Namib Plain, and (3) the Tsondab Planation Surface (Ollier 1977). The first two have been discussed and provide only the most tenuous connection. The last is no better despite being a prominent planar unconformity between the dune sands of the Sossus Sand Formation and the underlying Tsondab Sandstone Formation. It cannot be identified unequivocally north of the Kuiseb River.

Nor are there recognizable remnants of the Tsondab Sandstone Formation on the north side of the Kuiseb valley, although the Formation and perhaps the Tsondab Planation Surface probably extended well north of the Kuiseb canyon. What little evidence is available suggests that the Tsondab Sandstone Formation thins northward (Besler and Marker 1979; Besler 1984) and thus may have lensed out north of the canyon. Fluvial episodes that fashioned the Tsondab Planation Surface across the Tsondab Formation may then have removed such remnants on the north side of the valley—as they obliterated much of the Tsondab Sandstone along its eastern boundary.

Distinct remnants of a planation surface can be seen cut across marbles on lower slopes of the Hamilton and Witpoortberg ranges, south and north of the study area. These remnants may be the northernmost equivalents of the Tsondab Planation Surface. Hüser (1976) documented rock-cut surfaces on the flanks of marble ridges in the Ubib River drainage to the east, as did Beaudet and Michel (1978) against schist and granite outcrops in the Inner Namib. Equivalents in the Tumas basin remain speculative, however. It is probable that topographic smoothing, erosional and depositional, occurred when the Tsondab Planation Surface was current.

Assignment of the Namib Platform landforms to the erosion surface scheme of Lester King is equivocal. The gravel plains of the study area have been ascribed to the post-African surface by King (1967), and by Besler (1984) to the Tsondab Planation Surface. Ollier (1977) has seen the Tsondab Planation Surface as the probable equivalent of King's (1976) Moorland (African) surface. Without naming them, Spreitzer (1965-66) identified four surfaces in the Central Namib at the approximate altitudes of 200, 400, 600, and 900 m, the first two of which appear to be represented in the study area.

Chapter 4

CENOZOIC CLIMATES

. . . pleistocene, pliocene, miocene, are regrettable barbarisms.

—H. W. Fowler, *A Dictionary of Modern English Usage*

The Tertiary to Mid-Pliocene Times

Three views on the nature of more ancient climates of the southwest coast of Africa, especially on the origin of the Namib Desert, have been held successively. From early in this century researchers have generally followed Kaiser and Beetz, whose documentation of deep weathering profiles and karstification ("Karrenbildung, Verkarstung, Dolinen") led them to propose that late Cretaceous land surfaces experienced higher rainfall and a humid climate ("warscheinlich humides Klima," "höherer Niederschläge") (Kaiser 1926:435), which subsequently became distinctly desertic (Beetz 1926). Koch's entomological studies (Koch 1961, 1962) appeared to confirm this view. He concluded from the diversity and endemism especially of the tenebrionid beetle fauna, that long periods of desertic environment were implied. Koch was responsible for the Namib becoming known as one of the oldest deserts in the world.

The late 1970s brought a reinterpretation of extant faunal material, the discovery of fossiliferous beds in the southern Namib (Corvinus and Hendey 1978) and especially a series of reinterpretations based on deep-sea core data. These promoted awareness of the phenomenon of upwelling as a cause of aridity and suggested significantly later initiation of drying. Van Zinderen Bakker (1975) reasoned that aridity in the Namib may have arisen during Oligocene times when cold water conceivably became more common offshore as a result of the cooling of circum-Antarctic deep water to present temperatures (Shackleton and Kennett 1975).

Siesser (1978) argued, however, that aridification probably began with strong upwelling in the late Miocene, 25 million years after the early Oligocene cooling of bottom waters. The evidence of faunal remains, along the southwestern Cape coast especially, led Tankard and Rogers (1978) to suggest a Pliocene onset of aridity, intensifying in the Quaternary. Indeed, most authorities have agreed that the southwest coast of Africa was significantly wetter during some or most of the Tertiary, on the strength of marine and terrestrial evidence.

Thus Proto-Decima et al. (1978) envisaged the existence of subtropical environments along the coast based on the evidence of planktonic foraminifera and thermophilous nannofossils during the Paleogene. Dingle et al. (1983) suggested that coastal Paleogene floras from Namibia to Mozambique were evergreen, with palm forests and summer rainfall. They concluded that the wide coastal plains associated with the Oligocene regression "must have been well vegetated" with a "relatively warm and wet climate compared to that of the present-day west coast" (Dingle et al. 1983:313). Although marked aridification was seen as coinciding with strong Benguela Current upwelling during the early Upper Miocene, these authors interpreted earlier (early Oligocene to mid-Miocene) increases in *Coccolithus* phosphorous content to indicate "weak but spasmodic" upwelling with corresponding decreases in rainfall; this change they interpreted as leading to the replacement of tropical by "subtropical-temperate floras" in the Oligocene (Dingle et al. 1983:312).

Interestingly, Paleogene terrestrial floras at Banke in Namaqualand incorporate *Podocarpus*, as well as a protea-like genus and xerophilous leaf types (Estes 1978), suggesting to Tankard and Rogers (1978:322) the existence of an "ecotonal area between temperate rainforest and sclerophyllous vegetation, with a relatively dry climate and summer rainfall," although significantly wetter than today. Deacon (1983) has questioned whether this evidence is indeed indicative of dryness.

Data on the Miocene has become moderately prolific: marine vertebrates, microfossils, macro invertebrates and sedimentological evidence suggest warmer hydroclimates, and terrestrial floras and vertebrates suggest greater effective moisture than present. Proto-Decima et al. (1978) suggested a continuation of Paleogene oceanic conditions until early Upper Miocene times when significant cooling occurred. Doubled sedimentation rates, significant increases in organic carbon and diatom frustule abundance, and a constant rise in marine organism phosphorous content mark a major period of cooling as a result of upwelling (Siesser 1978).

Diester-Haas and Schrader (1979) documented significant fluctuations toward present conditions in terms of changes in diatom and radiolarian abundance, lowered ratios of planktonic to benthic foraminifera, an increase in fish debris, and the appearance of phosphate grains. Within this generally cooling trend, but by contrast with present conditions, several different mollusk taxa (Tankard 1974a) and three warm-water shark species (Hendey 1981) from the Varswater Formation Gravel Member are strongly suggestive of coastal temperatures 3-5 Centigrade degrees warmer than those of today. Cosmopolitan mollusk communities also indicate expanded margins of tropical oceanic water (Tankard and Rogers 1978).

Vertebrate fossils from several coastal sites (browsing ruminants, proboscideans, and a crocodile) suggested to Hopwood (1929) the existence of a savanna vegetation association, and to Dingle et al. (1983:315) that "the coastal zone of Namibia at least, was warm, well-watered, and well-vegetated in the early Miocene." On this evidence Deacon suggested "a woodland vegetation of trees and grassland" (in Dingle et al. 1983:315). From mammalian assemblages in the southwest Cape, Hendey (1981) proposed monsoonal climates with heavy summer rainfall, the horse *Hipparion* indicating open woodland and savanna.

Pollen analyses show the existence of palms along the west Cape coast (Coetzee 1978) in the late Oligocene and mid-Miocene, and macroremains of the tree *Curtisia dentata* suggest that a temperate rainforest was much more widespread during the Miocene than it is today (Tankard and Rogers 1978). Deacon (1983) concluded that fynbos-like vegetation with wet winters came into being only in the Pliocene, when pollen evidence shows the dominance of modern vegetation types (Coetzee 1978). Dingle et al. (1983) noted that this change to shrub associations does not necessarily imply a significant rearrangement of atmospheric circulation patterns, since the fynbos probably evolved under a summer rainfall regime.

Dingle et al. (1983:316) concluded from this mass of evidence that "the disappearance of extensive woodlands in the southwest Cape and Namaqualand in early Pliocene times, and their replacement by grassland and ultimately by fynbos or shrubland in Upper Pliocene and Pleistocene times," was a trajectory controlled by the "cooler and progressively more arid conditions that obtained as Benguela upwelling became established on a large scale." Pliocene faunal remains nevertheless suggest that environments were still unlike those of today: Hendey (1976) envisaged grasslands—with horse and alcelaphine antelope—crossed by riverine woodlands with giraffids and other large herbivores. Tankard (1974b) interpreted the presence of the grass-eating rhinoceros *Ceratotherium* in Namaqualand to indicate the existence of grasslands during the Pliocene.

Noting that these interpretations are at variance with the evidence of the Tertiary geology of the Namib Desert, Ward et al. (1983) have advanced a third and opposite perspective, one which they suggest subsumes the faunal and floral evidence. They interpreted a variety of widespread Tertiary sediments and soils to indicate dominantly arid or semiarid climates in preupwelling times. More than a dozen sedimentary bodies—some unreported in the extensive reviews of subcontinental geology and paleoenvironments by either Tankard and Rogers (1978) or Dingle et al. (1983)—include eolian deposits, conglomerates, and fluvial sands with massive, well-developed calcretes, all of which probably fall between the early Paleogene and mid-Miocene. All the depositional environments are consistent with arid and semiarid settings (Ward et al. 1983). These authors consequently argued that the faunal material does not necessarily indicate coastal savannas in preupwelling times because of the recently documented existence, in the driest parts of Namib, of many large species of savanna mammal. Such "noteworthy anomalies" (Ward et al. 1983:181) call into question previous paleoenvironmental interpretations of the faunal assemblages.

Pedogenic calcium carbonate accumulations in the Central Namib have attracted attention for many years. The Kamberg Calcrete Formation and similar pedogenic carbonates are suggestive of the wettest environments documented by any of the major lithologies of the Central Namib. These calcretes extend west from the Great Escarpment into the study area and are generally thought to imply precipitation markedly higher than that of today (Logan 1960; Hüser 1976; Besler 1977; Blümel 1979, 1982); Blümel (1976) suggested optimal figures of 600 mm/yr., though lower minimum figures have been proposed by Goudie (1973). The massive Kamberg Calcrete Formation, which crops out prominently along the south rim of the Kuiseb canyon (Ward 1984), is possibly late Miocene in age (Ward et al. 1983). It is probably a pedogenic accumulation, ascribed a formative rainfall regime of 400-450 mm/yr. by Yaalon and Ward (1982).

Ward et al. (1983) thus rightly call attention to influences other than the Benguela Current per se as the aridifying agent. They envisage both the Benguela Current and the subsequent associated upwelling as intensifying prior arid or semiarid conditions. The distinction between the more localized effect of the semipermanent subtropical anticyclones is often not appreciated, however. Periods of consistently warmer hydroclimates in the Tertiary may have sufficiently decreased downward heat flux to have promoted increased low level atmospheric instability and rainfall along the coast—perhaps to levels found today in the Inner Namib or Khomas Hochland. Increased precipitation and insolation (as a result of reduced fog-bank development) may well have allowed semiarid, tropical Thorn Woodland of the Inner Namib to advance westward to the coast, eclipsing the present perarid and superarid desert associations. How far inland the effects of warmer hydroclimates may have been felt remains an open question.

Even assuming that ribbons of riverine forest vegetation continued to occupy major river beds, it remains difficult to reconcile the many flora- and fauna-based interpretations with those of the geological record, by casting nonriverine areas as drier semiarid savannas rather than, for example, forests of significantly moister climates.

Much Pleistocene change is ascribed to the forcing effect of the Benguela Current. Difficulties remain in understanding the dynamic complexities of the current and their connections with past environments. For example, Tankard and Rogers (1978) have argued that upwelling occurred along the southwest Cape coast in the early Pleistocene, whereas the abundance of the thermophilous oyster *Striostrea margaritacea* further north indicates that upwelling did not occur on the Namaqualand or Namib coasts at that time. These authors have commented on the Pleistocene effects of interaction of the Benguela and Angola currents. Extension of the Benguela Current north of the equator (CLIMAP 1976) and the cooling of its core by 4-5 Centigrade degrees (Moreley and Hays 1979) may document an opposite (glacial?) extreme. Allanson (1984) has proposed that a recent "warm event" in the current may have originated in warmer Brazilian water.

Upper Pliocene to Mid-Pleistocene

In this period the the paleoenvironmental evidence changes to include mainly geomorphological and sedimentological phenomena, some now dated radiometrically. Most of the evidence discussed below is poorly dated, however, and its assignment to the end-Cenozoic is meant as a rough indication of age only. Frustratingly little floral or faunal data exists, but marine fauna and archeological residues have given some information on the more recent past. Against the background of aridity and semiaridity, well documented for this time span, many indicators suggest fluctuations in the moist direction.

Compared with other lines of paleoenvironmental evidence, fluvial data is widespread and better dated. Hüser (1976) presented an eloquent picture of fluvial and pedogenic activity in the escarpment zone. In the Ubib River drainage southeast of the Tumas basin (figure 1), three phases of valley widening, punctuated by carbonate duricrusting, have been identified (Hüser 1976). Hüser (1976) reasoned that valley widening and transport of coarse debris across slopes were related to increased rainfall and runoff, and that periods of calcification indicated relatively lower precipitation. Conditions as dry as those of today are specifically excluded. Marker (1979) and Lancaster (1984) documented phases of incision and deposition in the endoreic Tsondab drainage (figure 1), all of which suggest variably stronger discharge. Former terminal playa sediments downstream of present river end points thus imply progressive long-term aridification in the Namib (Lancaster 1984), especially where major sand dunes transgress these features. Selby et al. (1979) dated the Narabeb lacustrine deposits, 38 km west of the present Tsondab River playa (figure 1), between 210,000 and 240,000 years B.P.

The continued existence of such deposits is further taken to indicate the lack of major flows as far as Narabeb since that time. The Oswater Conglomerate, possibly Early to Middle Pleistocene in age, is likewise considered to represent greater energy levels in the Kuiseb River than exist today (Ward 1984).

Three geomorphic events characterize the evolution of the Lower Terrace in the Ugab River valley 200 km north of the Tumas basin; these events have been interpreted as evidence for Pleistocene climatic variations, from "semiarid" to "extreme arid" (Mabbutt 1952:361). The fining-upward sequence of fluvial deposits in the Uis River, a tributary of the Ugab River, is taken to indicate declining Quaternary rainfall in the Central Namib Desert (Korn and Martin 1957; Ward et al. 1983). Generally unremarked is the paleoenvironmental import of the Uis River alluvial tin deposits that Korn and Martin (1957) assigned to climates conducive at least to seasonal stream flow.

Interdigitated eolian sand lenses in the Oswater Conglomerate of the Kuiseb River canyon imply that the present longitudinal dunes of the Sossus Sand Formation were in existence during deposition of the conglomerate. Ward et al. (1983) suggested that the present linear dune field south of the Kuiseb River originated in a Pliocene arid phase. Hüser (1976) also argued for increased aridity in the eastern Namib from the existence of thick blown-sand accumulations (Tsondab Sandstone Formation?), which underlie the calcrete crusts in the Ubib River drainage, where dune deposits are today less prominent.

Support for ideas of greater precipitation during the Pleistocene come from Besler's (1976) morphometric and textural analyses of the above-mentioned linear dunes in the Namib Sand Sea. Besler (1976) considered that low-angled slopes of the basal plinth, and differences in grain texture between plinths and mobile dune crests, are indicative of fluvial activity deep within the dune sea; the fluvial episode intervened between an earlier eolian phase represented by plinth sands and a later phase represented by the mobile crests. Lancaster (1981a, 1983) has suggested that the textural differences are eolian-derived phenomena rather than an eolian-fluvial distinction.

The continued existence to present times of the Miocene Kamberg Calcrete may indicate the lack of significant moisture in the Central Namib since formation of the calcrete (Ward et al. 1983), although Marker (1982) identified widespread karstification of this unit in the Inner Namib. Marker (1982) argued consequently that significantly wetter climates are implied, at least against the escarpment. Environments responsible for karstification were apparently of sufficiently short duration to prevent wholesale destruction of the Calcrete. Ward (1984:459) identified tufa deposits in the Kuiseb River canyon (Hudaob Tufa Formation), the existence of which suggest "locally wetter conditions." Brain (1985) documented larger tufa deposits at several points along the escarpment to the east. Although formative environments of tufa development remain a matter of controversy (Partridge 1985), Butzer et al. (1978) have argued that the major Ghaap Escarpment tufa carapaces of the northern Cape Province imply semiarid to subhumid past climates.

Influxes of fresh water diatoms and phytoliths from Middle-Upper Pliocene and Pleistocene offshore sediments have suggested increased wind velocities, the highest Pleistocene opal values being correlated with most intense upwelling and probably higher wind velocities (Diester-Haas and Schrader 1979). Sea-surface temperature fluctuations are apparent, minimum temperatures being recorded during the Pleistocene south of the Walvis Ridge (Embley and Moreley 1980). Tankard and Rogers (1978) have argued for a warmer, nonupwelling Benguela Current at some time in the early Pleistocene for coasts north of the southwest Cape. Diester-Haas (in Lancaster 1984) has stated that eolian inputs dominate the nonmarine fractions in core sediments from the Walvis Ridge during the last 200,000 years. These findings support pollen data from the same cores, pollens characteristic of modern vegetation associations (van Zinderen Bakker 1984a).

In contrast to the views of Ward et al. (1983) favoring an early origin of the Namib desert, Axelrod and Raven (1978:112) have interpreted the renowned thermophilous "oyster line" occurrence near Oranjemund—of probable early Pleistocene age (Haughton 1932; Carrington and Kensley 1969; Tankard, in Axelrod and Raven 1978)—to indicate not only that "the Benguela Current was not as strong or as cold as at present," but that "the desert and mediterranean climates [of southern Africa] were not then in existence as pronounced regional features."

Korn and Martin (1957:14) reported "abundant Late Chelles-Acheul tools" from calcified river terraces in open sites distant from modern sources of water. The existence of both tools and terrace were interpreted as indicating significantly more mesic environments. Amelioration of aridity during the Pleistocene is implied by stone tools of Early and Middle Stone Age (ESA, MSA) within the dune sea (Sandelowsky 1977). The extreme paucity of terminal Acheulian sites (Korn and Martin 1957) suggests a period of intervening aridity. ESA tools cluster around the paleolake at Narabeb (Seely and Sandelowsky 1974), the above-mentioned, former end-point playa of the Tsondab River 38 km west of the present playa. ESA tools at the archaeological site of Namib IV, southwest of Gobabeb in the Dune Namib (figure 1), show affinities with East African assemblages dated between 700,000 and 400,000 B.P. (Shackley 1980), a period confirmed by fossils of a later representative of *Elephas recki*. This evidence led Shackley (1980:341) to posit the existence of a "small ephemeral lake" and "savannah grassland with sufficient fodder for elephants."

Upper Pleistocene and Holocene

Korn and Martin (1957:19) stated unequivocally that "the existence of pluvial conditions . . . is . . . proved by the widespread occurrence of MSA implements in the driest parts of the Namib desert," with a trend in later MSA times toward location at water supply points. The distribution of MSA artifacts—unrolled tools on higher terraces and rolled within the major incision of the Tsondab valley—suggest that this phase of incision occurred during MSA times (Lancaster 1984). Evidence of Late Stone Age (LSA) tools is absent from the dune sea, suggesting the existence of inhospitable conditions during the late Last Glacial and Holocene (Sandelowsky 1977).

Relict lacustrine sediments have been described from several points in the Namib Desert. A succession of four units with diatomaceous, organic, and clay components have been documented by Eriksson (1978) in the Koa River valley of Namaqualand. Uranium/thorium dates suggest emplacement in early Upper Pleistocene times, and again around 40,000 B.P. and 8000 B.P., although the earliest may be Middle Pleistocene in age (Butzer 1984). Undated diatomaceous lake sediments have been documented from Kannikwa near the Namaqualand coast (Kent 1947; Kent and Gribnitz 1985).

More than 150 radiocarbon determinations have been derived from sediments within and surrounding the Kuiseb River canyon zone, and from pedogenic calcrete, calcified root casts, and flood silts on the flats and in valleys south of the Kuiseb River (Vogel and Visser 1981; Blümel 1982; Heine and Geyh 1984). Despite the number of dates, interpretations have differed, partly as a result of site-specific differences of interpretation, and partly as a result of differing emphases placed on dates related to the Kuiseb River canyon zone, in which hinterland moisture fluctuations have impinged. Indeed, Vogel (1982) and Rust et al. (1984) have stressed the importance of distinguishing between sample sites located within the areas that have been affected by Kuiseb River flow versus those that are indicators of conditions autochthonous to the Namib proper. Dates in the Kuiseb valley provide a more complex history of combined local and hinterland effects.

Thirty-seven of the earlier dates have been interpreted as indicating a moist period in the Namib before ca. 19,000 B.P. Vogel (1982) and Rust et al. (1984) have judged that rainfall was higher before 28,000 B.P. on the strength of carbonate precipitation on the desert flats above the altitude of past courses of the Kuiseb River. "The last time more moist conditions prevailed in the Namib Desert ended about 28,000 B.P. Subsequent to this date fluctuations of the rainfall intensity occurred above the escarpment without drastically changing climatic conditions in the desert" (Vogel 1982:208). The cessation of travertine deposition in a small cave near Rössing led Heine and Geyh (1984:468) to suggest that "drastic" drying set in somewhat later at 25,000 B.P. Rust et al. (1984) have contested this interpretation of the travertine dates.

Possibly assuming that significant rainfall inland acted to increase precipitation in the desert as well, Deacon and Lancaster (1984:68) have suggested that the Namib (rather than simply the lower Kuiseb valley) "was considerably wetter than today" from 40,000 B.P. or earlier, to ca. 19,000 B.P.

Of the 154 radiocarbon dates derived from various workers, only three occupy the nearly 6000-year gap between 18,000 and ca. 12,000 B.P. Deacon and Lancaster (1984) ascribed this notable hiatus to a lack of carbonate deposition consequent upon marked aridification and possibly increased windiness. The view that the Pleniglacial in the Namib Desert was arid (Wieneke and Rust 1975; Rust and Schmidt 1981) holds only latterly according to the evidence of radiocarbon data clusters. Rust et al. (1984) promoted the argument, from evidence of two glacis terraces that cut the Homeb silt deposits (Rust and Wieneke 1974) and from unclustered Pleniglacial dates, that the Pleniglacial was indeed moist in the Namib and not arid as Heine (1982) has suggested. Van Zinderen Bakker (1984b) has concluded, however, on the strength of pollen studies of Sossus Vlei silts, that no increase of local rains has occurred since 18,000 years ago.

Successions of dates related to runoff in the Kuiseb River have been summarized by Butzer (1984:50), who stated that "the Namib dates indicate protracted river discharge, with extensive ponding as well as soil carbonate mobilization by throughflow water between 33,000 and 18,000 B.P. (with a possible interruption about 26,000 B.P.)."

Other authorities have suggested that this moist period ended earlier with a dry spell spanning the millennia before ca. 19,000 B.P. (Vogel 1982). Vogel's opinion relies on the interpretation of the stratified silts at Homeb (figure 1) as playa sediments laid down during dry-phase recession of the Kuiseb River end point—in the style of the smaller Namib Desert rivers (Marker and Müller 1978; Rust and Wieneke 1974). Butzer (1984) has suggested that the silts are best understood as the product of aggradation under more humid conditions.

Vogel (1982) implied that some time after 19,000 B.P. an increase in rainfall allowed the Kuiseb River to incise the 30-m-thick silts. Certainly there is agreement that by 14,000-13,000 B.P. smaller rivers in the Namib (Tsauchab and Tsondab) had begun to flow, depositing silts in presently nonfunctional parts of their drainages. These silts have provided five dates. Root casts at the coastal lagoons of Meob and Conception bays have given another four dates (corrected) between 13,000 and 8640 B.P. (Vogel and Visser 1981; Vogel 1982). These are interpreted as indicating the calcification of roots of dense stands of reeds during the progressive decline of near-surface water tables (Vogel 1982; Butzer 1984; Deacon and Lancaster 1984). The cessation of pedogenic carbonate accumulation in the Etosha Pan region in northern Namibia has been documented at a similar time (ca. 9200 B.P. by Rust 1985).

Evidence for Holocene environments shows possibly three moist fluctuations, based mainly on dated archaeological residues. Occupation levels at Mirabib rock shelter (80 km south of the study area) suggest moister conditions from prior to a radiocarbon age of 8400-8200 B.P. (Sandelowsky 1977; Vogel and Visser 1981). A laminated caliche horizon has yielded a date of 8385 B.P., indicating enhanced soil water movement (Blümel 1982). A second is strongly suggested between 6840 and 5200 B.P. by fully nine dates, five from Mirabib rock shelter, two from Cha-Ré shelter 80 km east of Mirabib, and one each from shelters in the Brandberg massif and Zwei Schneider 60 km north of the Brandberg (Vogel and Visser 1981).

A third moist fluctuation is as well documented as the former two, although it is generally unremarked. Ten dates from rock shelters in the Brandberg and Erongo mountains fall between 3268 and 2240 B.P. (Martin and Mason

1954; Beaumont and Vogel 1972; Vogel and Visser 1981). The absence of dune-dwelling moles from a dated owl pellet assemblage at Mirabib indicated to Brain and Brain (1977) that dunes had not existed in the vicinity during the last 8000 years.

Conclusion

Whereas the sequence of paleoclimatic events is becoming clearer for the late Pleistocene and Holocene, this is not the case for earlier periods. The biological evidence for the Paleogene, though often indirect and reliant on interpretations of the effect of increasing upwelling, is replete with suggestions of climates permanently and significantly moister than present. By contrast, the geological evidence consistently points to climates no moister than semiarid in the Namib Desert for tens of millions of years. Because of distance and the discontinuous behavior of the Benguela Current, it may not be valid to draw paleoclimatic conclusions for the Namib coast from faunal data in Namaqualand and the southwestern Cape Province.

Fluvial sediments with interdigitated dune sand, the progressive distal shift of river end points, and the age of pedogenic accumulations indicative of subhumid-semiarid conditions all attest to aridification of the Central Namib from the late Miocene onward, perhaps under the influence of permanently established upwelling in the Benguela Current.

Reports of caliche in the Ubib River valley (Hüser 1976) fit this trend, but the interpretation of much increased moisture to explain the incision phases seems out of step with the mass of other evidence, as does Marker's (1982) report of karstification of the calcretes and Korn and Martin's (1957) report of tin deposits with implied seasonal river flow.

The pattern of later climates is characterized by oscillation, with apparently alternating phases of habitation and abandonment of playas in the dune sea, and the alternation of periods of high water table with climates like those of today in and around the lower Kuiseb River.

General circulation models for Namib paleoclimates are based on manipulation of modern climate controls, namely the position and intensity of the subcontinental and South Atlantic anticyclones, and on the cold Benguela Current. The origin of upwelling, as a comparatively well-defined point of change in the long-term evolution of Atlantic Ocean circulation, has received attention as a controlling variable. As such, evidence for the appearance, strengthening, weakening, and migration of the current, as a control of coastal aridity at minimum, has been much discussed. At the Pleistocene time scale, upwelling has been regarded as partner in the control of moist excursions of climate against the background of existing desertic conditions.

The central players in theories of Pleistocene climatic change, however, are the westerly wind belt and tropical circulations. Van Zinderen Bakker (1967,

1976) long held the view that northward migrations of cyclonic storm tracks are the cause of phases of increased moisture as far north in the Namib Desert as the latitude of Walvis Bay, since this latitude is the most northerly for recorded cyclonic storms. Conversely, during the early Holocene, excursions of the Intertropical Convergence Zone were viewed as bringing precipitation to points as far south as Walvis Bay.

More recently, van Zinderen Bakker (1980) suggested that strengthened circulation during glacial times allowed the penetration of temperate disturbances into the Namib Desert not necessarily accompanied by a northward shift of wind belts. More recently yet, from his own palynological research at Sossus Vlei in the Namib dune sea (van Zinderen Bakker 1984b), and from studies on photosynthetic pathways from bone in the southern Namib Desert (Vogel 1983), van Zinderen Bakker (1984b) has suggested that cyclonic rainfall may not have reached the latitudes of the Central Namib during the Last Glacial Maximum.

Models favoring penetration of tropical systems have been suggested to explain phases of increased precipitation in central southern Africa as far west as the Namib Desert (Butzer et al. 1978; Rust et al. 1984). Cockcroft et al. (1987) applied Tyson's model (Tyson 1986) of southern Africa's long-term wet and dry spells to dated climatic trends in the late Pleistocene and Holocene. The model proposes distinct reversals in tropical and temperate windfields and particularly east-west shifts in a major standing wave of the upper tropospheric circulation, the latter a novel element in analyses of subcontinental paleoclimates. "The model predicts an increase in winter rainfall over an expanded winter-rainfall region at the time of generally lowered temperatures during the late Quaternary. It accounts adequately for the poleward spread of higher rainfall after 9000 BP—undoubtedly a summer rainfall phenomenon" (Cockcroft et al. 1987:176). In this model, the boundary between regions dominated by tropical and temperate systems in full-glacial times is poorly defined in Namibia, but the winter rainfall region appears to have expanded further north in Namibia than elsewhere in the subcontinent, quite possibly as far north as the study area. Models for the Outer Namib are complicated by possibly independent behavior of the Benguela Current: weakening of the current certainly suggests that seaboard precipitation would rise, as Butzer et al. (1978) have suggested, to levels that characterize the plateau further east. Interpretations of the evidence are conflicting, however. The zone of upwelling in Pleniglacial times is variously reported to have moved north (CLIMAP 1976), to have remained unchanged (Rust et al. 1984), and to have increased under the influence of greater wind speeds (Moreley and Hays 1979; Newell et al. 1981).

Chapter 5

LITHOSTRATIGRAPHY, ARCHITECTURE, AND PALEOENVIRONMENTS OF NAMIB GROUP DEPOSITS IN THE LOWER TUMAS BASIN

So strong indeed is the tendency toward transformation [of stream behavior] that it is only in the few streams of permanent [water] supply . . . that definite channels are maintained.

—W J McGee, *Sheetflood Erosion*

Aggrading alluvial plains in shield deserts are poorly understood geomorphically and in terms of the sediments they host. Analysis of the Tumas units is based on an appreciation of depositional controls at different scales of aggregation, in particular at the scales of lithofacies, sediment-body shape, and the alluvial plain. It is necessary to introduce background information and terminology for the discussion of scale, lithofacies, and sediment body architecture. Based on major attributes of the Tumas basin presented in chapters 5 and 6, a general model of arid sedimentation in the setting of a confined valley is outlined at the end of the next chapter.

Lithofacies and Architectural Elements

Lithofacies

Miall (1977, 1978) has erected a widely applied lithofacies scheme of approximately twenty generalized types, a scheme designed to encompass "most kinds of fluvial deposit" (Miall 1984:139), and those fluvial units in deltaic sequences (table 2). Lithofacies mnemonics (codes) refer to primary textural categories (G: gravel; S: sand; F: silt/clay), each with subdivisions indicating "distinctive texture or structure of each lithofacies" (Miall 1984:139). Three subdivisions for closely related eolian sands are included, and Eyles et al. (1984) have extended the system to include all diamict facies (D: conglomeratic facies; S: sandy clastless diamictic facies; table 2). The purpose of facies schemes is to promote "judicious simplification" and to facilitate interpretation (Miall 1984:139); as such, they have been adopted for this study.

TABLE 2.
Lithofacies classification
(after Miall 1978; Eyles, Eyles, and Miall 1984)

Facies Code	Lithofacies	Sedimentary structures	Interpretation
Gms	massive, matrix-supported gravel	grading	debris flow deposits
Gm	massive or crudely bedded gravel	horizontal bedding, imbrication	longitudinal bars, lag and sieve deposits
Gt	gravel, stratified	trough crossbeds	minor channel fills
Gp	gravel, stratified	planar crossbeds	linguoid bars or deltoid growths from older bar remnants
St	sand, medium to very coarse, may be pebbly	solitary or grouped trough crossbeds	dunes (lower flow regime)
Sp	sand, medium to very coarse, may be pebbly	solitary or grouped planar crossbeds	linguoid, transverse bars, sand waves (lower flow regime)
Sr	sand, very fine to coarse	ripple marks of all types	ripples (lower flow regime)
Sh	sand, very fine to very coarse, may be pebbly	horizontal lamination, parting or streaming lineation	planar bed flow (lower and upper flow regimes)
Sl	sand, fine	low-angle crossbeds (<10°)	scour fills, crevasse splays, antidunes
Se	erosional scours with intraclasts	crude crossbedding	scour fills
Ss	sand, fine to coarse, may be pebbly	broad shallow scours	scour fills
Fl	sand, silt, mud with intraclasts	fine lamination, very small ripples	overbank or waning flood deposits
Fsc	silt, mud	laminated to massive	backswamp deposits
Fcf	mud with intraclasts	massive with molluscan fauna	backswamp pond deposits
Fm	mud, silt with intraclasts	massive, desiccation cracks	overbank or drape deposits
Fr	silt, mud	rootlets	seatearth
C	coal, carbonaceous mud	plants, mud films	swamp deposits
P	carbonate	pedogenic features	soil

(TABLE 2 *continued*)

Diamictic lithofacies		
Diamict: D		
Dm	matrix supported	structureless mud/sand/pebble admixture
Dc	clast supported	(as above)
Dcm		massive
Dcs		stratified
Dcg		graded
Sands: S		
Sm	massive	
Sg	graded	
Sd	soft sediment deformation	
Fine-grained (mud): F		
Fm	massive	

Architectural Elements

Although fluvial sedimentary environments are probably the most widely explored and intimately understood of all sedimentary environments (Allen 1965; Harms et al. 1975), Friend et al. (1979) have observed that "little is known about the processes forming the two- or three-dimensional geometries of fluvial sediment bodies" (p. 39); they note that such geometries hold importance for considerations of mineral exploration, tectonics, and paleoclimates. It is appropriate therefore, to examine questions of sediment body shape, especially in relation to geomorphology and climate,as an aid to analysis of Namib Group deposits.

Potter (1967:340) has emphasized that definitions of depositional environments should be based on the "geomorphic concept": he states that "a sedimentary environment is defined by a set of values of physical and chemical variables *that correspond to a geomorphic unit of stated size and shape*" (emphasis added). In the following discussion an attempt is made to investigate the relationship between geomorphic environment and sediment body shape, both because of its theoretical interest, and because the three-dimensional architecture of the Tumas Formation sediments is readily ascertained and thus highly tractable to this new mode of analysis.

Friend (1983) and Miall (1985, 1987), among others, have recently argued the importance of understanding the three-dimensional configuration, or "external geometry" (Miall 1985:268), of fluvial sediment bodies. Miall (1985, 1987) has noted that significant problems pertain to interpretations based on facies models alone: analysis is restricted to vertical profiles, that is, to one di-

TABLE 3.
Architectural elements of fluvial sediments
(after Miall 1985)

Element	Symbol	Principal lithofacies assemblage
Channels	CH	any combination
Gravel bars and bedforms	GB	Gm, Gp, Gt
Sandy bedforms	SB	St, Sp, Sh, Sl, Sr, Se, Ss
Foreset macroforms	FM	St, Sp, Sh, Sl, Sr, Se, Ss
Lateral accretion deposits	LA	St, Sp, Sh, Sl, Sr, Se, Ss; less commonly Gm, Gt, Gp
Sediment gravity flows	SG	Gm, Gms
Laminated sand sheets	LS	Sh, Sl; minor St, Sp, Sr
Overbank fines	OF	Fm, Fl

mension; only smaller features such as bedforms and gravel bars are dealt with; and the convergence of forms by different processes reduces the diagnostic value of the forms.

Miall (1985) has accordingly proposed a "new method of analysis" (p. 261) involving a flexible hierarchy of "architectural elements"—a term taken from Allen (1983)—which, he suggests, are limited to "only about eight" in number (p. 268). These are common fluvial macroform structures of variable size, but as much as "a few hundreds of meters in width and length" (p. 268) (table 3 and figure 9).

In Miall's scheme, the architectural elements are hierarchized according to the significance of bounding surface so that one type may contain within it smaller types. The largest element is always the CH (channel element), which envelopes the other seven.

Applied with proper caution, the architectural element scheme is designed to facilitate the reconstruction of depositional environments as the "summary of an environment" (Miall 1985:300). Typical combinations of these elements are suggested in a group of twelve major fluvial depositional environments—termed "river models" (table 4)—a group that Miall (1985) stresses is not exhaustive. Thus, for example, proximal alluvial fans can be typified by the elements GB (gravel bedform), SG (sediment gravity flow), and minor SB (sandy bedform elements). Alluvial fans and proximal outwash braidplains commonly exhibit GB and minor SB elements only (models 1 and 2, table 4), the GB element comprising, for example, gravel sheets and bars (Gm lithofacies), crossbedded gravel bedforms (Gp lithofacies), and scour fills (Ge and Gt lithofacies). An advantage of the models is that they can accomodate the great variability of fluvial deposits because many combinations of the "constant" architectural elements are possible (Miall 1985).

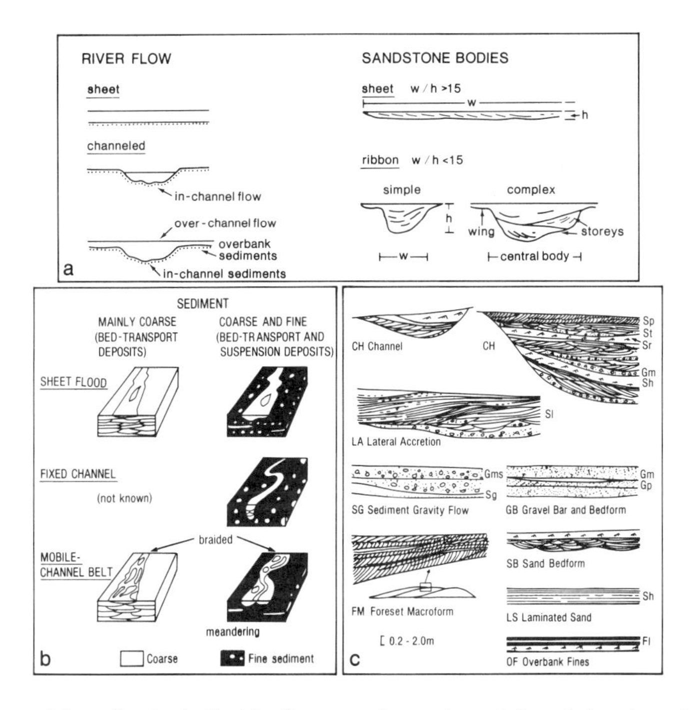

Fig. 9. Degrees of channelization in fluvial sediments. a. External morphology: body, wings, sheets (after Friend et al. 1979). b. Stream behavior and fluvial architecture (after Friend 1983). c. Fluvial architectural elements (Miall 1985, with permission).

The scheme is peculiarly appropriate also to studies of late Cenozoic sediments, since external geometry can often be ascertained far better than it can for many ancient rock bodies. Furthermore, in the case of younger sediment bodies, formative geomorphic controls often remain extant in the landscape.

Geomorphic Units: Architecture and Climate

Miall (1985) specifically avoided discussion of tectonic and climatic controls of architectural elements and river models because he foresaw that "a considerable amount of new field work and rethinking within an 'element' framework will be necessary before significant advances can be made" (p. 298). In this study an attempt is made to promote such rethinking in the following ways: (1) by suggesting a new architectural element because of the inadequacy of the CH element as a descriptor of fluvial architecture in some environments; and (2) by investigating the effect on fluvial architecture of climate, one of several controls of sedimentation external to the fluvial system. Climate is classed as a control of allocyclic type, with tectonics, eustatics, sediment type, and initial paleoslope (Reading 1986).

TABLE 4.
River models
(after Miall 1985)

	Stream model no.	Sinuosity	Sediment type	Characteristic elements
1.	Alluvial fan with abundant debris flows	low	gravel, minor sand	GB, SG (SB)
2.	Alluvial fan with few debris flows	low	gravel, minor sand	GB (SB)
3.	Larger gravel-bed stream (some large alluvial fans)	low to interm.	gravel, minor sand, fines	GB, SB (OF)
4.	Gravelly river (flat-topped bars)	high	gravel, minor sand, fines	GB, LA, OF (SB)
5.	Coarse-grained meandering stream	interm. to high	sand, pebbly, minor fines	SB, LA, OF (GB)
6.	Classical sandy meandering stream	high	sand, minor fines	LA, SB, OF
7.	High sinuosity suspended-load stream	high	fine sand, silt, mud	LA, SB, OF
8.	High sinuosity anastomosing, stable channel stream	low to high	sand, fines	SB, OF (LA)
9.	Broad, shallow, abundant sand bedload stream	low to interm.	sand	SB, FM
10.	as no. 9, with greater bar-bar top facies differentiation	low to interm.	sand, minor fines	FM, SB, OF
11.	Distal braidplain with shallow channels	low	sand, minor fines	SB (OF)
12.	as no. 11, with very flashy discharge	low	sand, minor fines	LS (OF)

Geomorphic Units and Sedimentary Environments: Scale Considerations

As a basis for discussion both in this chapter and ensuing appreciations of the Tumas plain and arid plains in general, it is necessary to clarify the question of scale of the phenomena to be discussed, especially since some of the features are relatively uncommon. Scale considerations are given little attention or are confused in much sedimentological writing.

Potter (1967) considers size a crucial characteristic of the geomorphic unit within which sedimentation takes place. Scale has been treated mainly as a duality. Beerbower's (1964) distinction between internal and external controls of sedimentation has been widely used as an analytical construct, especially as applied to the choice of controls that are appropriate to the analysis of features and areas of different size. "[C]ycles [of sedimentation] induced by allocyclic (extrinsic) mechanisms should be basin wide in scale; autocyclic mechanisms will produce cycles of limited extent—similar in scale to the individual depositional environments within an alluvial plain system" (Beerbower 1964:1846). These distinctions have become known as "internal/intrinsic" and "external/extrinsic" in the geomorphic literature (Schumm 1981). Miall (1984) has encouraged the use of two scales of analysis in sedimentology, the "facies-scale" and "depositional system-scale," the former broadly related to autocyclic mechanisms, the latter to allocyclic mechanisms (table 4.6 in Miall 1984).

Schumm and Lichty (1965) added complexity to this bipolar scheme by proposing a continuum of controlling variables. The dependence, independence, and even relevance of variables change depending on the time span within which any specific geomorphic system operates. Schumm and Lichty's (1965:118) landmark theoretical statement applies to these issues: "The distinction between cause and effect among geomorphic variables varies with size of a landscape and with time . . . the more specific we become the shorter is the time span with which we deal and the smaller is the space we can consider." This statement can be interpreted to show how controls comprise a hierarchy or continuum rather than merely a bipolar analytical approach. More important, it encourages consideration of the effect of extrinsic controls on relatively small fluvial subsystems; and it shows that the nature of the variables can change as the time span of the analysis is changed.

For example, sand sheets can evolve specifically by the interaction of an extrinsic and an intrinsic control—in this case, rates of basinal subsidence and stream avulsion (Allen 1978; Bridge and Leeder 1979; Blakey and Gubitosa 1984). The operation of climate, another extrinsic variable, at the broadly facies scale, is the topic of this introductory section on sediment architecture.

The scales of analysis—termed "levels" in this section—are derived from a selection of the more overtly geomorphic and better known models. The models illustrate the disparate scales and nature of geomorphic units that have been employed in attempts to understand fluvial depositional environments.

This attempt at classification is preliminary and is not meant to be exclusive. It is designed more to aid in conceptualizing the varied scales that seem to be important in the functioning of fluvial depositional systems.

At the lower end of the hierarchy, the first four levels are those outlined above and derived from Miall (1984, 1985). Level 1 is the lithofacies scale. Level 2 comprises the majority of architectural elements separated by mainly second order discontinuities. Level 3 represents the larger architectural elements bounded by third and fourth-order surfaces (Miall 1985). Features of levels 1, 2, and 3 all appropriately fall at the facies scale of analysis: Miall (1985) has even described the largest architectural element as "autocyclic" (p. 272). Miall's (1985) river models comprise level 4.

Higher levels are suggested in the literature. Level 5 can be defined as typifying primary divisions within a depository. Rust's (1978) basin subdivisions—the one-dimensional "confined valley" and the unconfined (that is, two-dimensional) situation of the alluvial fan[1]—are examples typical of level 5 environments. Friend (1978) and Rust and Gibling (1990) have argued the importance of valley setting to facies disposition: Friend (1978) showed that facies frequency can be controlled by changes in valley width. Allen's (1965) four alluvial facies models (alluvial fan, meander-belt flood plain, flood plain with straight river, flood plain with braided river) deal with restricted reaches of flood plains. These are level 5 features. "Constriction convexities," referred to in chapter 6, are features that arise in the Namib Desert in valley-confined flood plains and fit appropriately at this level in the hierarchy.

Rust's (1978) third subdivision is the alluvial plain of significantly larger size, a level 6 feature. It can be thought of as the typical basin-scale environment ("depositional system-scale") of Miall (1984:202). Schumm's (1977, 1981) idealized drainage basin (as opposed to sedimentary basin) model is divided into a zone of coastal deposition (zone 3) fed with sediment by inland zones of erosion and transportation (zones 1 and 2). Zone 3 represents a generalized depository at this level of aggregation. Galloway and Hobday's (1983) "ideal fluvial system" (p. 73) is more elaborate and is directed at larger basins than Schumm's (1977) comparatively small basin. In effect it contains two basins of different type, a montane basin with aggrading river plain, arbitrarily draining into a separated prograding coastal plain crossed by a single river. Allen's (1965) "alluvial facies associations" (meandering trunk river with lateral fans; major alluvial plain with avulsing river) fit best at this level.

Attempting a simple geomorphic classification of sedimentary basins, Miall (1981) has identified nine basin models controlled by trunk and tributary stream geometry (transverse and longitudinal), delta type, and tectonic or distal environment (marine, lacustrine).

Large multiple depositories characterize level 7 environments. Allen (1965) modeled a multiple-river coastal plain in which several major rivers contribute to a sediment body, usually of large proportions, assigned here to level 7. Such environments are closely interrelated with neighboring marine depositional environments, and their size prescribes significant isostatic feedbacks that influence sediment body shape. The great sediment wedge along the north coast of the Gulf of Mexico comprises interlinked sedimentary lobes arranged around six major river axes (Galloway 1981). Large basins of inland drainage

---

[1] In an intriguing excercise in geomorphology and sedimentology, recent interest in fan-deltas has focused on discrimination of deltas, alluvial fans, outwash plains, braidplains, and braided rivers. Confinement of some of these by neighboring fans or valley walls appears to be a major control of depositional style and architecture (McPherson et al. 1987; Nemec and Steel (eds.) 1988; Rachocki 1981).

such as the Kalahari depression of southern Africa, supplied by multiple sources of sediment, are of this order.

The levels identified above coincide approximately with orders 9 (ripples) to 3 (medium-scale geological units) of a geomorphic hierarchy outlined by Chorley, Schumm, and Sugden (1984). Although higher orders of landform correspond with increasing spans of time and shifts in the suite of other controlling variables (Schumm and Lichty 1965), climate remains a crucial variable. Subsequent arguments suggest that climatic effects can impinge sufficiently strongly on fluvial architecture so that reconstruction of architecture may aid in revealing formative climates.

SH: Sheet Architectural Element

To adequately address the problems of architectural description and analysis of sediment bodies in the Tumas basin, it is proposed that another element be added to Miall's (1985) eight. Termed the SH: sheet macroform, its peculiar interest here is that it has implications concerning climate, an allocyclic variable, although Miall's (1985) scheme, as noted, is conceived of as the product of controls mainly internal to fluvial systems: CH elements—the largest in his scheme—are described as "autocyclic" (Miall 1985:272).

Definitions. Sheetlike sediment bodies have been described from various sedimentary environments and lithologies, but especially with respect to sand bodies because of the economic potential associated with sheet sandstones. With specific reference to fluvial sandstones, Friend et al. (1979) have used a ratio of 15 to distinguish between ribbons and sheets. Blakey and Gubitosa (1984) further distinguished narrow sheets from broad sheets (ratios 15-100 versus >100 respectively). Sheets may be simple or compound, and in the latter case, they may be multistorey and multilateral (see Potter 1967, and Friend et al. 1979, for the origins of these terms). For convenience, I shall use the latter descriptive terms and width/thickness ratios as definitions of the SH: sheet element.

It is a measure of the preliminary status of these analyses that ratios are the criterion of sheet classification rather than scale of geomorphic environment of the type suggested by the levels above.

Sheetlike forms are well attested in the recent and ancient geological records. Miall (1985) lists several elements in recent deposits that can be sheetlike in external form, though all are subsumed within the CH form. At higher levels, sheetlike forms are familiar from alluvial fans (e.g., Bull 1972, 1977) and modern, confined valley settings (McKee et al. 1967).

Present classification of sheet forms tends to mix scales considerably: they have been classed as (1) channelized (confined by valleys) and simple (e.g., Collinson 1978) or compound (e.g., Friend et al. 1979); and (2) unchannelized (unconfined by valley walls) and simple or compound (Friend 1978). Multiple sheets more than 100 km wide (Campbell 1976), and some even greater than 150 km wide (Blakey and Gubitosa 1984), have been referred to by Miall (1987:5) as features of the "regional" architecture. These belong to geomorphic settings of levels 6 and 7. The scale distinction between single-event sheets—which equate with Miall's (1985) second-order elements, specifically the SG: gravity flow type—and multiple sheets implies quite different geomorphic environments as between level 2/3 architectural elements and level 6 alluvial plains.

Since sheetlike entities are known from features at the architectural element levels 2 and 3, it seems necessary, if only at a descriptive level, to propose the existence of an SH: sheet architectural element, particularly at the level of higher-order bounding surfaces, that is, at the level of Miall's (1985) CH: channel element. The terms "SH: sheet" and "SH element" are accordingly used here in Miall's (1985) sense for features at levels 2 and 3.

Analytical Significance of the SH: Sheet Element. Sheetlike macroforms appear to hold importance not only descriptively, but also at an analytical level. At lower levels such as the facies model scale, the primary analytical distinction lies between channelized and unchannelized flow: Friend (1983:349) has stated that "recognition of channels is a major step in the analysis of an ancient fluvial formation," since it distinguishes fluvial deposits related to channels of mobile and fixed behavior from those related to unchannelized stream behavior.

Channelized flow can generate sheet forms by lateral migration of mobile river belts, and by close-spaced avulsion of fixed channels (McGee 1897; Collinson 1978; Tunbridge 1981, 1984; Friend 1983). Both modes of sedimentation generate lateral accretion and stacked, or multistorey architecture. On the other hand, sheets always appear to result from unchannelized sheetflood behavior (Friend et al. 1979; Friend 1983) (figure 9). Just as large CH elements encapsulate smaller CH and other elements (Miall 1985), this discussion shows that SH elements can likewise contain CH, SH, or other elements nested within them.

Although Miall (1985:275) defined the CH form to include "almost unconfined, sheet-like channels," and "practically imperceptible channel margins, sloping a few degrees or less," unchannelized, convex-upward types such as SG: gravity flows typical of alluvial fans fit poorly into a scheme in which all architectural elements are defined as subordinate to the CH: channel form. It may be argued that some of Miall's (1985) smaller architectural elements allow for sheetlike morphologies (e.g., LS, SG, LA, SB) as an optional characteristic. These are sheet forms that develop in channels (Miall 1985), and they can be distinguished from unchannelized types.

The foregoing arguments suggest that sheetlike morphology needs substantially greater recognition. In light of the analytical importance that appears to attach to the existence of sheet forms, it may be useful to categorize architecture into primary divisions of CH: channel and SH: sheet. In general, there appears to have been resistance to the full analysis of sheets as phenomena uncongenial to traditional concerns of fluvial sedimentology. Unchannelized flow is accommodated poorly in most reconstructions of fluvial environments. Collinson (1978:584) concluded an extended argument on the genesis of sheet sands thus: "When channel margins and lateral accretion bedding are absent, the safest policy at present is to question whether so-called 'coarse members' thinner than 1-2 meters need be equated with channel deposition at all." The study of fluvial architecture evidently remains in its infancy.

Namib Group Sediments in the Lower Tumas Basin

The Tumas and Leeukop Formations are those bodies of sediment centrally located in the study area,[2] the former being uraniferous and exposed at surface, the latter nonuraniferous and investigated only by subsurface probe. The former, in particular, has yielded much evidence for late Tertiary environments in the Central Namib.

Leeukop Conglomerate Formation

The Leeukop Conglomerate Formation is defined as those sediments that underlie the Tumas Sandstone Formation.[3] As such it is probably significantly longer than the known 36 km length of the Tumas Formation (A-D, figure 4), and reaches a maximum thickness of ca. 90 m in the valley center as judged from the two deepest probes. The Leeukop Conglomerate is documented from only seven diamond drill holes clustered in the central study area. Fourteen deep (50 m) percussion drill holes on the southern Gawib Flats and fifty from sector A-B probably penetrate the top of the Formation, but the description here is taken from the diamond drill cores in which stratigraphic relationships and lithofacies are not obscured. The Leeukop Formation is defined as filling the entire 3-5 km width of the buried valley and thus undoubtedly comprises sediments derived locally and from further afield. The walls of the buried valley constitute the lower boundary; the upper boundary is inferred from lithological changes determined from drill samples, and crudely coincides with a bedrock shoulder between 20 and 35 m below surface, suggesting a period of valley widening.

---

2 Descriptions cohere with various requirements of the International Subcommission on Stratigraphic Classification (1976), the North American Commission on Stratigraphic Nomenclature (1983), and the South African Committee for Stratigraphy (1980).

3 The name is derived from the prominent marble hill situated on the southern watershed of the Tumas basin; it is one of the few names to appear on maps of the area. The cores are deposited in the Anglo American Base Camp at Husab.

The bedrock valley of the Tumas basin in the lower sector A-B widens (figure 10a) such that the valley sides are composed of young Namib Group sedimentaries, presumably of the Leeukop Conglomerate Formation. The latter is known only from drill logs, however, there being a notable lack of outcrop owing to flatness of the topography and well-developed, surficial gypsic soils. The Leeukop Conglomerate Formation undoubtedly fills buried tributary valleys that enter on the north side. As such, the Formation comprises one of the largest noncoastal deposits in the Tumas area.

Lithofacies. Close visual examination of the diamond drill cores and analysis of drill logs indicate twenty-two beds in the reference section (one of the two deepest diamond drill holes, no. 7/13[4]— figure 11). Neighboring holes (750 and 1500 m to the north, and a third 3 km west) show no unequivocal stratigraphic correspondence, indicating that discontinuous strata are characteristic of the Leeukop Conglomerate. Unit thicknesses vary: the average is 2.7 m, the range 30 cm to 14 m.

The Leeukop Conglomerate Formation is dominated by fine to coarse pebbly conglomerates in a matrix of mainly coarse to very coarse sand, although three units display a multimodal distribution of pebbles, sand, and silt/clay (figure 11). The remaining better-sorted beds comprise fine, medium, and coarse sands. Two beds only displayed cobble-grade clasts.

A spectrum of hues occurs, but most are of low chroma and high value (all color determinations according to the Munsell Color Company 1954 notation, on dry samples), generally gray, but they range from reds through pinks to white as a result of impregnation by calcium carbonate. Twenty units responded to the application of hydrochloric acid, some strongly. Most units are correspondingly consolidated to cemented, with only one unit weakly coherent and one other noncoherent as a result of apparent openwork gravel fabric. Bedding characteristics, difficult to ascertain in drill cores, were seldom apparent.

Evaluation. A facies assemblage dominated by undoubtedly discontinuous gravelly units and located within the confines of a paleovalley, suggests that the Leeukop Conglomerate Formation was deposited in a fluvial setting, probably by a braided river. Poorly developed or nonexistent cyclicity is another feature of types of braided stream deposition (e.g., Miall 1977, 1984).

The consistently conglomeratic nature of the Leeukop Conglomerate Formation, the high percentage of carbonate-rich beds, the lack of gypsum impregnation, and its relatively great thickness, are significant facts in evaluating changing styles of sedimentation in the lower Tumas valley. Little more can be said other than that the Formation comprises mainly Gm lithofacies and therefore probably accords with the GB: gravel bedform element.

---

[4] Longitude coordinates are given first.

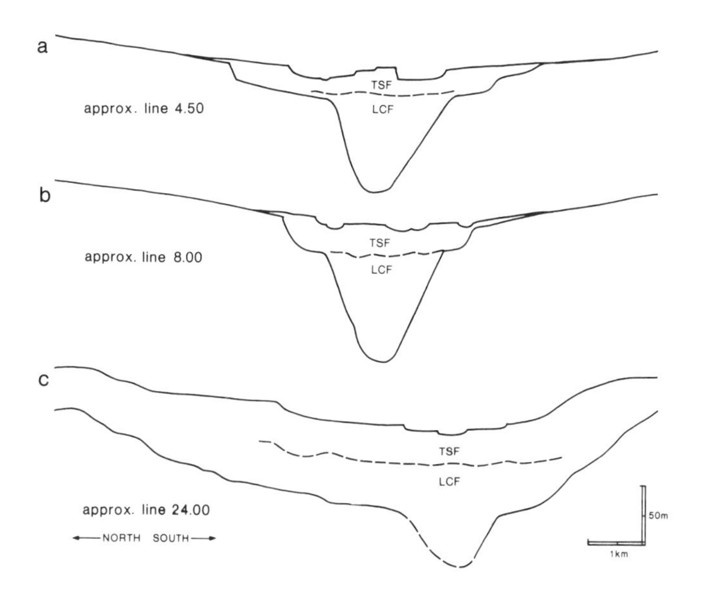

Fig. 10. Tumas canyon (north-south sections) illustrating Leeukop Conglomerate Fm., Tumas Sandstone Fm, and buried canyon-wall shoulder at contact. a. Section in sector A-B (fig. 4). b. and c. Sections in sector B-C. Vertical exaggeration 25x.

Tumas Sandstone Formation

The Tumas Sandstone Formation comprises two Members, a 10-35 m-thick red sandstone overlain by a thin (2-2.5 m) but ubiquitous gray gravel and gravelly sand (figure 12). Heavy impregnation by gypcrete in the upper few meters of the Formation is a process that has given rise to pseudo-bedding plane development and vertical joints, which occasionally cut both Members (figure 13a, b). The reference section is pit 6.50/13.170. Much variability exists with respect to the size and position of minor gravel lenses, logged and unlogged. More than nine hundred percussion drill holes and one hundred backhoe trenches penetrate the Formation and provide the cross-sectional data.

Member 1 Red Sandstone. Member 1 comprises a massive red sandstone within which are intercalated volumetrically minor facies of coarse gravel and very minor clay stringers. Member 1 has been identified sporadically in drill holes as far as 60 km inland and to within 10 km of the coast. It occupies the 3-5 km-wide floor of the lower Tumas valley throughout this distance and thickens westward from averages of 10-15 m to 15-20 m with a maximum known thickness of 37 m in the westernmost drill line (through point A, figure 4). For this reason it seems very likely that the Tumas Sandstone Formation continues toward the coast. From its distribution across the entire width of the Tumas valley floor, this Member can be described as a broad sheet in terms of its ex-

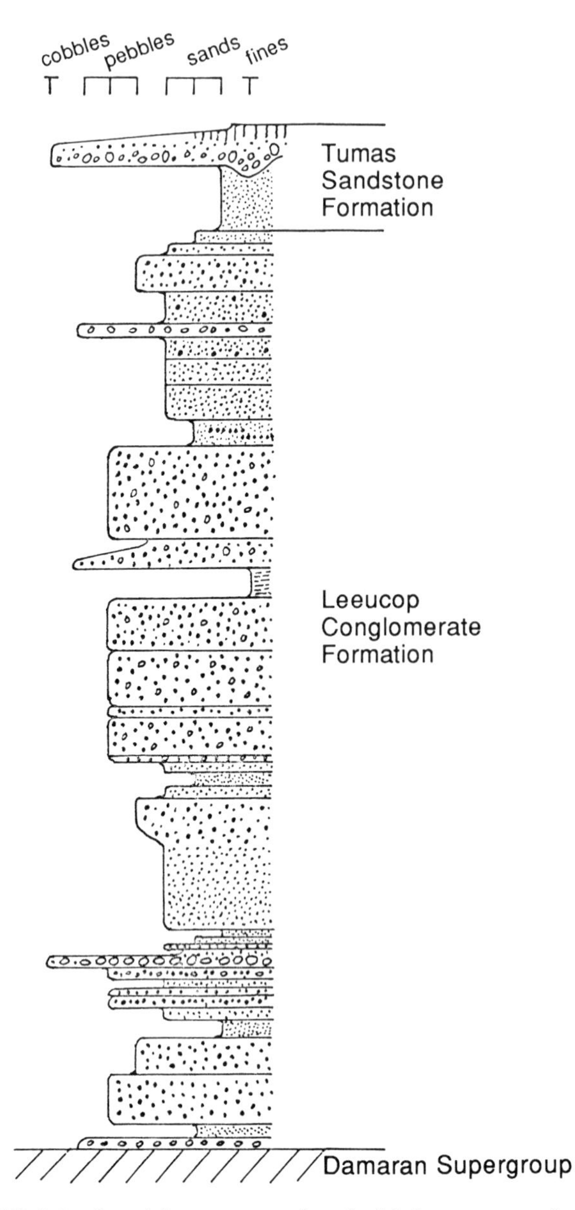

Fig. 11. Leeukop Conglomerate Fm. lithology (diamond drill hole 7.00/13.00).

ternal geometry (figure 10). Member 1 hosts anomalously high concentrations of supergene uranium. Although Member 1 crops out along tens of kilometers of stream bank, less disturbed lithologies were opened to view in almost one hundred backhoe pits and a half dozen trenches transverse to stream banks.

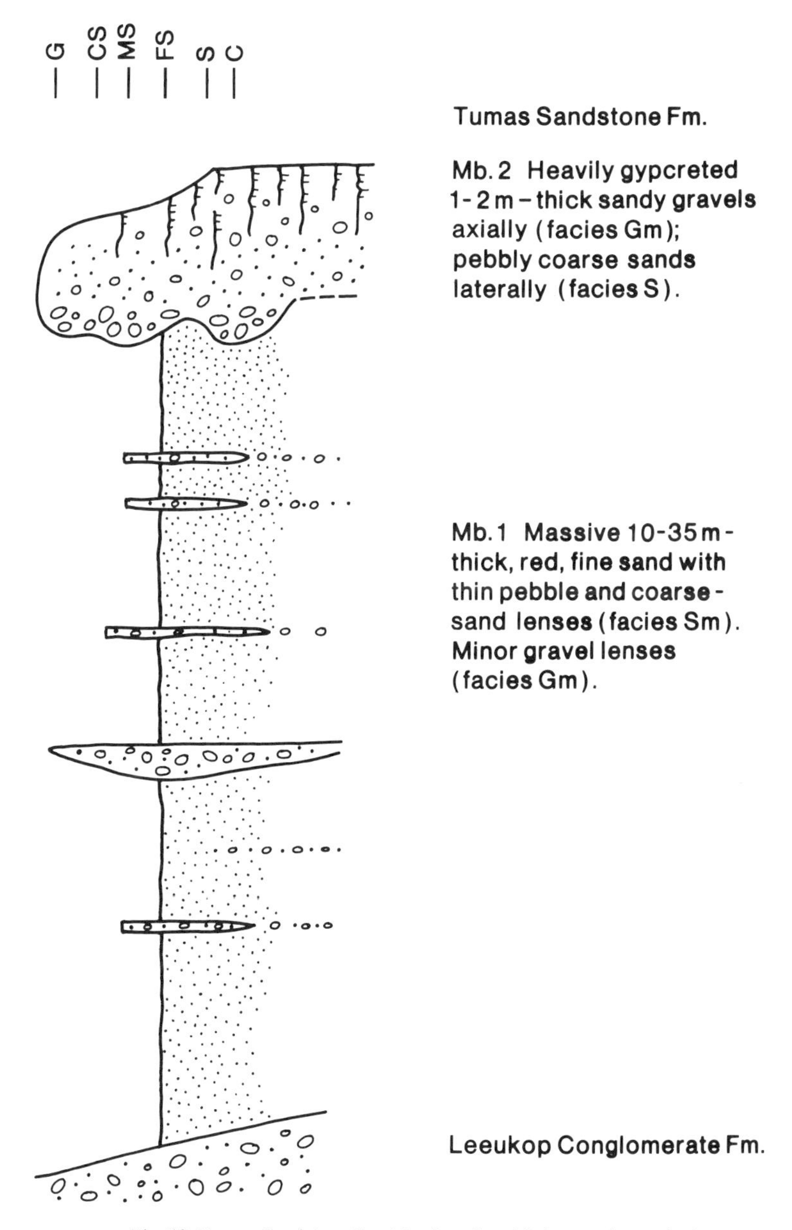

Fig. 12. Tumas Sandstone Fm. Members 1 and 2 (composite section).

Massive sands: facies Sm. The characteristically massive, poorly to very poorly sorted fine sand of this unit conforms with the sandy diamictic facies Sm of Eyles et al. (1984). Though predominantly consolidated, it is cemented in upper levels but sufficiently unconsolidated elsewhere, especially at deeper

a

b

Fig. 13. Tumas Sandstone Fm. Typical exposures in Tumas R. talweg of red sandstone (Member 1) overlain by grey-white sandy gravels (Member 2). a. Pseudo-bedding planes in Member 2; stream banks duricrusted by porcelaneous gypsum crusts; note sheetlike build of Member 2; surface S1 prominent (view upstream). b. Member 1 red sandstone; underlying pinker hues (backhoe dump) of transition zone between upper oxidized (red) and lower reduced horizons.

levels (with or without proximity to water tables), to promote caving of drill holes. Internal bedding is notably absent (figure 14), except in three localities: westward-dipping planar cross-bedded laminae occur in pits 24.75/22.50 and 6.32/11.50, and faint horizontal bedding is evident in pit 6.75/12.417. Higher-order bounding surfaces are no easier to find, an enigmatic fact in a Member known to thicken to at least 37 m. In particular, no channel forms were encountered. Highly discontinuous (1-3 m), horizontal, thin (<0.5 cm) lenses of coarser sand, with and without fine pebbles, occur very irregularly as the only consistent indication of bedding. Minor, discontinuous horizontal pebble stringers (of single pebble thickness), with diffuse lower boundaries, are suggestive of planes of bedding. Subhorizontal, near-surface, discontinuous to continuous veins of fibrous gypsum may have developed along original bedding planes, but may result equally from water table configuration. Horizontally aligned discoloration streaks (mm to cm thick) at 4-5 m depth in pit 7.75/12.375 are as equivocal.

Facies Sm varies laterally from a uniform red (2.5YR 4/6 to 2.5YR 5/8-4/8) to strong brown (7.5YR 5/6) (figure 14) beneath interfluve surfaces. Nearer drainage lines, colors are pinker (pink, 7.5YR 7/4; pinkish white, 7.5YR 8/2), lighter brown (light reddish brown, 5YR 6/4; light brown, 7.5YR 6/4) and yellower (yellowish red, 5YR 6/2; reddish yellow, 5YR 6/6; figures 13b, 14, 35). Many, prominent, coarse mottles (Soil Survey Staff 1975) also show undersaturation compared with colors more distant from drainage lines (pinkish gray, 5YR 6/2; light reddish brown, 5YR 6/3). Proximity to near-surface, gypsum-rich horizons lightens the characteristic reds and induces vertical color gradation.

A minor Fsc facies, encountered in only three pits (6.32/13.19, 6.50/12.93, 6.50/13.27), comprises consolidated brown (7.5YR 5/2) clay and silty clay, arranged in single, horizontal, thin (1-3 cm, reaching 5-6 cm) but usually massive bands, often wavy and sometimes contorted and slickensided (figure 15) as the result of abundant halite crystal growth. The beds are discontinuous between exploration pits 25 m apart in the area of close-spaced pitting.

Two bands separated vertically by 0.8 m are displayed in each of two pits (6.50/13.270 and 6.50/13.190). In a third pit (6.50/12.930), eight bands were documented 2.0 to 3.5 m below surface, distances between bands averaging 20 cm in this highly localized sample (figure 15). Two bands merge with one another in one instance.

Massive to crudely bedded gravels—facies Gm. Coarse sandy gravels of cobble grade, usually pebble supported and poorly sorted, are intersected in pits and drill cores as widespread but minor components of Member 1. The thickest unit (5 m), was encountered in pits in the valley center, but thicknesses are characteristically 0.5-1.5 m and almost exclusively massive. One pit (2.875/12.35) displayed openwork gravel at 5 m depth. Lenses encountered in the pits of the close-drilling area vary from <1 m in length to nearly 100 m in three

Fig. 14. Member 1 massive red sand (Sm facies); lower parts pink near talweg from gypsum impregnation. Pick marks 2-3 cm wide.

holes. Faint cross-beds dipping southwest in a 10-cm-thick set were encountered in pit 6.75/12.417. Matrix sands of this facies are often olive gray (5YR 5/2). Black manganese stain was noted in gravels of pit 2.875/12.35. Clast shape is heterogeneous (subangular to rounded), and consolidation varies from poor in openwork sections, to cemented. Occasional pockets of pebbly gravel, less than 1 m in diameter, with diffuse boundaries, are encountered.

Average thicknesses of 3-6 m are characteristic of the major lenses logged in drill holes. Units are seldom continuous between neighboring holes (250 m apart in the open-drilling area). GB: gravel bedform architecture is suggested.

Minor valley-margin gravels, evident from drill sections, interdigitate with Member 1 as lateral correlatives of this Member and may be the source material for some of the intra-Member 1 gravel facies.

Evaluation. The uniformly massive nature of a unit as large volumetrically as Member 1 relates to primary depositional environments. The horizontal, lenslike configuration of the minor gravel and clay/silt beds are primary bedding features, as are the discontinuous, horizontal lines of floating cobbles, pebbles, and coarser sand within the Sm sands. The clay bands even indicate that locally the sand beds were of the order of 1-2 m thick. The lack of obvious bedding in pits 7-10 m deep and in the diamond drill cores, and the existence of uncontorted bands of coarser material, preclude the possibility that all vestiges of first- and second-order bounding surfaces were postdepositionally obliterated. Such disruption has undoubtedly occurred in the upper 1-2 m, however.

It seems reasonable to conclude, therefore, that internal structures were poorly developed in Member 1 *ab initio*. Indeed, bedding disruption might be expected to parallel the patchy and discontinuous pattern of gypcreting, a pattern controlled first by depth, second by host caliber, and third by distance from stream courses. Its pervasiveness suggests therefore that massive bedding is a primary characteristic of the Member.

Interpretation of Sm lithofacies in a fluvial setting is not straightforward. The evidence adduced thus far, however, seems most consistent with a monotonous sequence of sediment gravity flows and a sediment source restricted to a narrow range of textures. The primary evidence for this interpretation is the massive and diamictic nature of the Member, features common to gravity flows of various types (Hampton 1975; Innes 1983; Schultz 1984). The lack of channelized units is another characteristic (Friend 1983), as is the phenomenon of thin, capping clay bands (Fsc facies)(Bull 1977).

The ubiquitous but very minor and poorly developed lenses of coarser, pebbly material require comment. Such lenses fall within the spectrum of rec-

Fig. 15. Member 1 massive red sand (Sm facies) with thin (1-6 cm) clay bands (arrows). Nail (top right) 3 cm long.

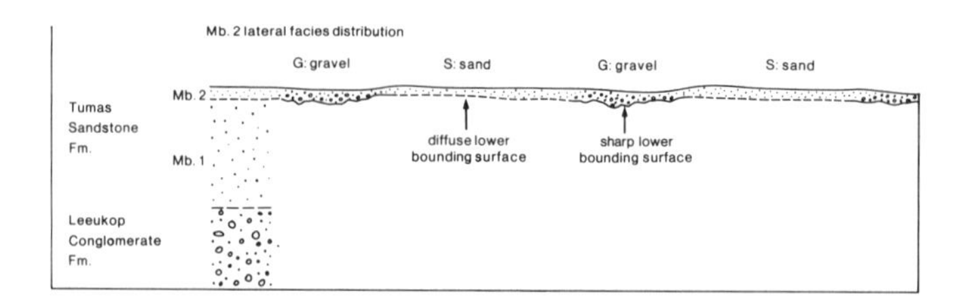

Fig. 16. Lateral gradation within Member 2: gravel-dominated facies axially, sand-dominated facies laterally (diagrammatic preincision section).

ognized gravity flow phenomena: Middleton and Hampton (1973) and Schultz (1984) have defined two major types of sediment gravity flow for those mass movements that lie in the continuum between fluids and plastics, namely fluidal flows and mass flows. The former include stream floods and turbidity currents, both of which give rise to load segregation as represented by the scattered lenses of traction load pebbles and sand.[5]

Hambleton-Jones et al. (1986:2276) report Minter's communication to them that the red sandstone comprises "a series of mass flow deposits."

The architecture can be described as SG: gravity flow elements (level 2) stacked horizontally and vertically. With the apparent entire lack of CH: channel elements, these are considered to comprise a sheet form (SH: sheet, level 3) and are undoubtedly sheetlike at valley-wide scales (levels 4 and 5).

Member 2 Gravelly Sands and Gravels. With an average total thickness of 2.0-2.5 m, Member 2 occupies most of the valley floor in the study area, allowing designation of the external geometry as a broad sheet. Its equivalents in other drainages comprise two units, as Martin (1950) noted, where best developed in valley-center locations. However, the distinction between the units is generally impossible to make in the study area.

A more salient aspect is the lateral change of texture. Ribbons of massive, mainly cobble conglomerates along drainage lines grade into widespread sand-dominated beds beneath the extensive, valley-bottom flats between the present-day channels (figure 16). The lower bounding surface of Member 2 is highly diffuse and planar where overlain by the dominant sandy units, but sharp and planar to wavy where gravel units are developed above it. The upper surface of Member 2 is modified by gypsum accumulation, subsequent microkarstification, and thin, sheetlike, patchy flows of essentially horizontally laid, coarse alluvial spreads, now often reduced to deflation lags.

---

5 There is little agreement on appropriate terminology for sediment gravity flow features: categories and terms employed by Middleton and Hampton (1973) and Schultz (1984) have been followed.

These two lithologies give rise to entirely different styles of gypsification, the gravelly sands to pervious mesocrystalline horizons, and the gravels to massive, alabastrine horizons. With gypsum accumulation well above 50% by volume, massive and pressure-structured gypsic and petrogypsic horizons (P) disrupt bedding and obscure texture and fabric in the upper 0.5-1.0 m of Member 2. A pseudo-bedding plane often separates the surficial petrogypsic and lower vein and nodular gypsums. Descriptions are taken from lower, undisturbed levels.

Member 2 is irregular in thickness in some places and regular in others, averaging 2.0-2.5 m in the study area, with a maximum known thickness of 3.5 m (pit 6.32/13.190). It lenses out toward the valley margins, exposing red sands of Member 1 at surface.

Lithofacies. Member 2 has constituent facies of three types. (1) S(P): sands and gravelly sands, and (2) Gp: planar bedded gravels, with associated Gm: massive bedded gravels, are all heavily impregnated with gypsum (P: soil lithofacies of Miall 1978). (3) Minor fine-grained units are facies Fl: laminated clay/silts, and Fm: massive clay/silts, developed very locally.

Facies S(P): sands and gravelly sands, display numerous pebbles, and occasional cobbles "float" in a mass of gypsum-rich sands in which structure is imposed by gypsum accumulation (pressure structures, honeycombs, and pseudo-bedding planes). Original bedding is entirely disaggregated. Sand

Fig. 17. Member 2 planar cross-bedded sandy gravels (Gp facies) with iron staining localized along reactivation surfaces and around some clasts (arrows).

grains are coarse to fine, often angular and often coated with grey and pink skins. Hues are varied and relate mainly to gypsum in its pristine (whiter) and weathered forms (pale yellow and brown grays). Along the lower contact, admixtures of Member 1 red sands impart light pink hues.

Gp and Gms facies are crudely bedded to massive, usually clast-supported sandy gravels (with lesser, matrix-supported and openwork lenses), which vary from pebble to boulder conglomerates, sometimes with rapid textural changes vertically and laterally, and sometimes with little change. Normal grading is characteristic where grading can be recognized, although reverse grading does occur. Lenses of coarse to medium sand are common. Clasts are predominantly vein quartz and quartzite, with gneiss much less common. Imbrication is occasionally evident. Angular to subangular cobbles of Member 1 sandstone are common in the lower meter, maximum clast size usually distinctly larger than surrounding gravel components.

Iron compounds impart orange and red stains to some reactivation surfaces in one locality (figure 17).

Minor, apparently massively bedded fine-grained units up to 50 cm thick have been noted in outcrop along stream walls in association with the Gms and Gp facies. These sediments are shot through with horizontal veins of fibrous gypsum, but their texture promotes a degree of consolidation that has prevented complete disaggregation. They display gypsum colors; very locally fine sands display a pale olive hue (5Y 6/3).

Five units of facies Fsc, laminated to thick-bedded clays and silts, 2 m in thickness, crop out very locally in the center of the study area in the wall of a major tributary (7.0-6.75/13.50, figure 18). Because of their fine texture, they are poorly gypcreted; laminae, mud drapes, desiccation cracks, and mud curls are evident.

A rare, but well-developed facies, Fr: rootlet-rich silt/clay and sandy silt, occurs in a few exposures in the eastern part of the study area. Abundant, vertically oriented, gypsumized root casts occur in a major pit (2.875/12.350) in association with extensive gypsum pressure structures and joints. These have developed to a depth of 3.5 m (figure 32).

Evaluation. Internal architecture consists of many sand-dominated SH: sheet elements and relatively few axial CH: channel and SH: sheet elements of the more gravelly units. Axial units are little thicker than the sandy facies with channel walls never apparent. Member 2 as a whole is developed across most of the alluvial plain surface so that it may be considered a broad sheet in terms of larger geometry.

Individual runoff events are interpreted as gravelly, shallow streamfloods along drainage lines with ephemeral banks; shallower discharges overflowed wide areas beyond the banks, entraining material from the underlying sandy Member 1. The geomorphic setting is reminiscent of Friend's (1978)

a

Fig. 18. Highly localized laminated to massive fine-grained sediments of Member 2 (exposure 6.75/13.50). a. Pseudo-bedding and joints generated by gypsum accumulation (upper half of exposure). b. Pressure structure and/or desiccation disruption.

b

braided, mobile channels and unchannelized slopes where sheet action operates. Some evidence of minor gravity flows in Member 2 is suggested by the Gms facies within the Gm gravels. The boundary between braided stream facies and gravity flows seems randomly crossed, although the latter are less common, suggesting within the gravel units a mode of behavior reminiscent of alluvial fans (Bull 1977); depositional style fluctuates between flows characterized by segregation of bed load and medium, to those characterized by distinctly viscous flow. Minor bodies of fine-grained sediments are consistent with local slack-water depositories on an alluvial plain. Clay bands are consistent with thin muddy carapaces, which characterize individual mass flows (Bull 1977).

Under the prevailing circumstances of facies disruption by heavy duricrusting, broad architecture (lack of channel features at the element scale and a broad sheet at the alluvial plain scale) may constitute a more secure basis for interpretation than disrupted facies sequences. On the basis of this and lateral facies type, it seems likely that the S: sand facies comprise sheet-flood and gravity-flow deposits.

The sharp erosional boundary between Members 1 and 2 beneath the Gm facies supports the notion of highly erosive, supercritical flow in gypcreted channel material; the diffuse boundary beneath the interfluvial sands suggests lower levels of gypcreting on the flats between major channels. Entrainment of indurated red sand cobbles, as intraclasts from Member 1, suggests that gypsification of the red sands was well advanced by the time Member 2 was emplaced. Olive green streaks suggest high water tables locally at topographic low points prior to gypsification.

Discussion

Depositional Setting of Sediment Gravity Flows

Although gravity flows are common in arid landscapes, those with the characteristics of Member 1 appear to be little reported. It has been mentioned that Member 1 comprises sediments ranging from fluidal flows and units overtly dominated by traction loads in minor lenses to an overwhelming majority (in terms of thickness) at the viscous end of the gravity flow continuum.

Geomorphic settings within which sediment gravity flows commonly develop have been classified variously. In his review Innes (1983) provided a "morphogenetic" classification, distinguishing flows in mountain valley bottoms, hillslope flows, lahars, and catastrophic flows. Innes (1983) also classed these phenomena roughly in terms of event magnitude. Bull (1977) identified three topographic environments conducive to gravity-flow development, namely alluvial fans, valley bottoms, and hillslopes.

The Tumas Formation gravity flows do not fit any of these categories well, although they display the characteristics of some: they are neither of the hillslope variety nor of the alluvial fan variety, since they lie within a broad alluvial plain of low declivity. Furthermore, alluvial fan debris flows typically decrease in thickness and abundance distally (Bull 1977; Hooke 1967), whereas the debris flow–dominated Member of the Tumas Sandstone thickens downstream and is generally of far greater size than typical flow-dominated units on fans. Neither is the valley-bottom category adequate, since the Tumas Formation flows have been preserved in great number, as in alluvial fans; this is in sharp distinction with Bull's (1977) opinion that valley-bottom flows are seldom preserved, since they are usually removed by subsequent river erosion.

It is necessary therefore, to refine concepts of depositional settings for sediment gravity flows.

Arid Alluvial Plains and Gravity Flows. Considering the extant alluvial plain of the lower Tumas basin, and the lack of mountain-front terrain necessary for the development of alluvial fans (Cooke and Warren 1973), it is suggested that the arid alluvial plain setting, as a depositional environment, accomodates the characteristics of Member 1. This setting is not well understood either geomorphically (Cooke and Warren 1973) or in terms of depositional environment (Blatt et al. 1980).

In order to evaluate further the Tumas Formation, it is necessary to propose attributes that might be included in a preliminary lithofacies and architectural model of alluvial plains in the Namib Desert.

Distal alluvial plains are preferred points of sediment accumulation and preservation in endoreic basins. Though less localized and possibly less efficient in terms of preservability than alluvial fans, plains are generally orders of magnitude larger than alluvial fans as depositories. In answer to Bull's (1977) perspective that valley floor flows have low preservation potential, it can be argued that there is no reason why sediment gravity flows, once generated, should not be preserved in arid environments, since endoreic basins lack the continuous reworking of through-flowing discharges that characterize exoreic basins; furthermore they hold all sediment supplied to them except for material deflated by wind.

Although it seems likely that gravity flows may be preserved in deserts, it is surprising that relatively few instances of diamictic fluvial units have been documented that are not interpreted as part of alluvial fans. Lithofacies descriptions from arid environments, both modern (Williams 1970, 1971; Karcz 1973; Picard and High 1973; Sneh 1983) and ancient (e.g., more recently, Horne 1975; Tunbridge 1981; Clemmensen and Abrahamsen 1983; Smoot 1983; Hubert and Mertz 1984), have become better known. The persistent lack of mention of diamictic (gravity flow) facies in depositional settings other than those interpreted to be alluvial fans is noteworthy; it suggests that gravity flow phenomena are not generated on alluvial plains.

It is submitted, however, that both the common lack of exposure of modern alluvial substrates at the distal ends of arid basins, combined with the minimal investigation of depositional loci in endoreic basins, may have militated against recognition of these features. Furthermore, though specific conditions of sediment supply may be necessary for their formation, these conditions are probably more often met than the paucity of reference suggests. Indeed, Blackwelder (1928) argued at length for recognition of gravity flows in desert landscapes of the American southwest, and gave many examples of mudflows projecting into basin-center playas. "Most desert playas are more or less completely margined by overlapping mudflows—a fact that has not been generally known." (Blackwelder 1928:474-5). Clemmensen and Abrahamsen (1983) mentioned regular interdigitation of debris flows and sheet floods with eolian beds in the Permian red beds of Arran.

Parallels with the Clarens Formation. Large numbers of preserved, terrestrial mass flows are characteristic of neither mountain valleys nor alluvial fans. But one African example at least holds striking parallels with Tumas Formation Member 1. The Triassic Clarens Formation of the Karoo Supergroup is a Formation which crops out along the borders of the Kingdom of Lesotho. Facies 3 of the Clarens Formation is a massive, fine sandstone, 32 m thick in one representative section (Eriksson 1981) underlying thousands of square kilometers in southern Africa. Bigarella (in Beukes 1970) interpreted the facies as sediment gravity flows generated in an arid environment. A subsequent study by Eriksson (1981:14-15) supported this interpretation of "mass flows" and "sandflows" emplaced "by mass movements of viscous debris flows which formed in arid environments when sediments became saturated during intense thunderstorms." Surrounding zones represent both wetter playa subenvironments and drier blown-sand subenvironments (Eriksson 1981). In the same Formation van Dijk (1978) documented a small lake encompassed by dunes, as well as graded sheet-flow units and possible ephemeral stream deposits. Further east, Facies 1 of the Clarens Formation has been interpreted as the product of fluvial redistribution of windblown sand both as channelized deposits, and as unconfined sheet-flood units deposited laterally from wadi courses (Eriksson 1979).

Although the lower Tumas basin is small by comparison with these basins and does not show the full array of arid subenvironments, Member 1 of the Tumas Sandstone is interpreted as analogous with Facies 3 of the Clarens Formation. Both salient characteristics of Member 1—dominance of fine sands and gravity-flow depositional style—are duly accounted for in this interpretation. McKee et al. (1967), Williams (1971), and Tunbridge (1981) all note that grain size in desert streams may reflect a sediment source control. "Sediment grade is not necessarily an indicator of stream power . . . source area composition may have a strong influence on the nature of sediment available for transportation, and only fine sand may be deposited from high-energy flows if no coarser material is available from the immediate source area" (Tunbridge 1981:90).

Supporting circumstantial evidence for these arguments comes from two facts: first, the modern dune field is located in the same area of the flood plain as the Member 1 dune field (figure 25) (figure 4, point C). Furthermore, the Tumas valley floor, downstream from point B (figure 4), is covered by a number of submodern gravity flows. Flow widths vary from 100 to 300 m and lengths seem not to exceed 3 km for individual flows. By comparison with Member 1, the flows are typically pebble- and cobble-rich and appear to be of the less viscous type. The phenomenon of unconfined, valley-bottom flows in the Tumas valley has been commented on briefly by Hövermann (1978).

It is envisaged, therefore, that a much expanded dune field was subjected to fluvial reworking to produce a long series of mass flows, mainly of dune-sand texture, with inevitable but minor admixtures of other textural classes. Poorly sorted flows were correspondingly emplaced as a sediment wedge down the width and length of the lower Tumas basin. Ribbons of gravel, of insignificant volume, were intercalated during particularly strong floods, but the great majority of events ended on the plain, where discharge ceased. The Tumas stream appears not to have reached the coast during the emplacement of Member 1 sands.

Two facts suggest that although stream activity dominated the Tumas basin in this stage, climates were still entirely arid. First, the existence of gravity flows strongly points to a geomorphic setting at the end point of a desert stream, where discharges become viscous and cease to flow. That is, stream loads were laid down within the basin rather than reaching the Atlantic Ocean or even the playa at the topgraphic low point of the basin. The great number of flows suggests that this situation persisted. Second, by arguments presented earlier in this chapter, the marked sheetlike architecture of Member 2 (that is, environments of levels 3 and 4), corroborates interpretations of aridity.

Paleoclimatic Implications

All past climates interpreted from the Leeukop Conglomerate and the Tumas Sandstone Formations fall within the ambit of aridity and semiaridity that appear to have characterized west coast Tertiary lithologies (Ward et al. 1983). Seven chronological stages are identifiable.

1. An early phase of exoreic fluvial activity excavates the buried Tumas and tributary canyons.
2. Conglomerates of the Leeukop Formation aggrade as an early, major depositional episode in the lower Tumas basin. Drill cores do not allow meaningful subdivision of the period of time that the Formation probably represents. The conglomeratic nature of the Formation and the discontinuous nature of individual beds are characteristic of

braided stream facies (Rust 1978). The 80-m-thick deposit may be associated with an early Tertiary transgression.

3. A hiatus in deposition is suggested by the phase of valley widening at about 20 m below surface in the lower Tumas basin. The shoulder is 0.5-1.0 km wide in some drill lines, expanding to 2 km downstream. The hiatus may encompass a long period, since the Tumas Sandstone is regarded as mid- to late-Tertiary in age.
4. Eolian deposition in the valley bottom and on flanking hillsides is implied by the volume of sands of dominantly eolian texture that make up Member 1 of the Tumas Sandstone. Sand was probably supplied from upstream by the Tumas River, much as it is to the Dune Namib by present ephemeral streams (Besler 1977, 1980, 1984). The size of the dune field was arguably at least as large as the area underlain by the Tumas Formation (>100 km^2), and far larger than the present small field (5 km^2) in the study area. A climate as dry as that of today is implied, dominant for a sufficient length of time to allow the dune field to expand. Considering that the present field may well represent Holocene accumulation in the small Tumas watershed, tens of thousands of years may be implied. Accepting the proxy nature of the evidence, the existence of large accumulations of dominantly blown-sand fractions point most obviously to a past, sand-covered landscape.
5. Subsequent fluvial activity is undoubtedly documented by the emplacement, mainly as gravity flows, of Member 1 of the Tumas Sandstone. A situation analogous to that of the Clarens Formation is suggested, namely that of dune sand redistributed fluvially during a distinct shift to wetter local climates. Expected admixtures of other particle populations were duly incorporated as a result of fluvial action, but fail to mask the nature of the original population of dune sand textures. Although the shift to wetter climates seems clear, the shift was not great. Autochthonous oxide reddening of sand grains may have commenced at this time and also indicates aridity; hardened oxide sand-grain coatings may relate to earlier periods of iron-compound mobilization.
6. Induration of Member 1 sands by gypsum suggests very tentatively a period of geomorphic quiescence, at least as arid as that of today. The well-defined contact between Members 1 and 2 and the indurated nature of Member-1 intraclasts at the base of Member 2 suggest that Member 1 may have been duricrusted prior to emplacement. Further from the drainage lines, however, the contact is less well defined and intraclasts are unknown, suggesting that, as today, duricrusting diminished with distance from stream beds.

7. Aggradation of the 2.5-m-thick Member 2 conglomerates documents a reversion to moister conditions, probably moister than those of Member 1 times. Sheet flooding, and boulder-grade loads, suggest runoff more vigorous than that during stage 5 or that of today. Such runoff probably reached the coast. Present stream flow barely penetrates the study area beyond the small dune field, whereas conglomerate units on terraces at Tumas Vlei, possible equivalents of Member 2, indicate that the stream has been substantially more active in the past. The Member 2 conglomerates appear to document a moister, perhaps semiarid, environment, in which deposition of sheetlike forms strongly supports the lithofacies interpretation.

The subsequent phases of supergene mineral emplacement, pedogenesis, and incision, subjects of the following chapters, indicate a trend to less active alluvial plain surfaces and reduced discharge in the Tumas basin.

Although unilluminating in paleoenvironmental terms, events 1-3 above are not inconsistent with the theme of Tertiary aridity adduced by Ward et al. (1983). The record of later stages is more detailed and indicates arid and semiarid environments, with an apparent distinction between an inferred eolian period and two stream-dominated episodes, the former weaker and the latter more vigorous. The minor, poorly documented, soil-forming episode 6 is the only overt evidence for morphostatic conditions in the generally aggradational period treated in this chapter.

Chapter 6

STREAM INCISION AND RELATED PHENOMENA

Rain . . . rarely falls in the vicinity of Walfisch Bay; but the gathering of heavy clouds in the eastern horizon every afternoon, and vivid flashes of lightning accompanied by distant thunder, clearly indicated that the interior of the country had been flooded. We soon had proof of this in the sudden appearance of the long-dormant Kuisip River, now swollen to an unusual height, [which] overflowed its banks, and threatened destruction to every thing that opposed its course.

—C. J. Andersson, *Lake Ngami*

By contrast with the issues of alluviation and sediment dispersal in the earlier periods of earth history in the Tumas area, this and succeeding chapters treat younger geomorphic events—phases of stream incision, soil formation and modern geomorphic activity. In particular, they encompass considerations of fluvial processes, development of various supergene minerals, and modern eolian effects.

The surfaces of the central lower Tumas basin and the duricrusts they support provide significant indications of past conditions in the arid core of the Namib Desert and are well developed only in the wide alluvial plain of the lower Tumas valley. Sequences of incision are not identifiable beyond the valley-bottom plain. Aggradational and erosional stream-profile geometries are analyzed. Intrinsic and extrinsic controls of these profiles are evaluated.

The analysis may also be described as the documentation of the behavior of the end point of an arid stream under changing circumstances of sedimentation and incision, an analysis that has not been undertaken in such detail for a Namib catchment.

The Tumas River stream bed is incised discontinuously. Apart from minor, discontinuous gullies cut 1-2 m into the calcreted surfaces of its headwaters, the Tumas River is incised significantly only in the study area,

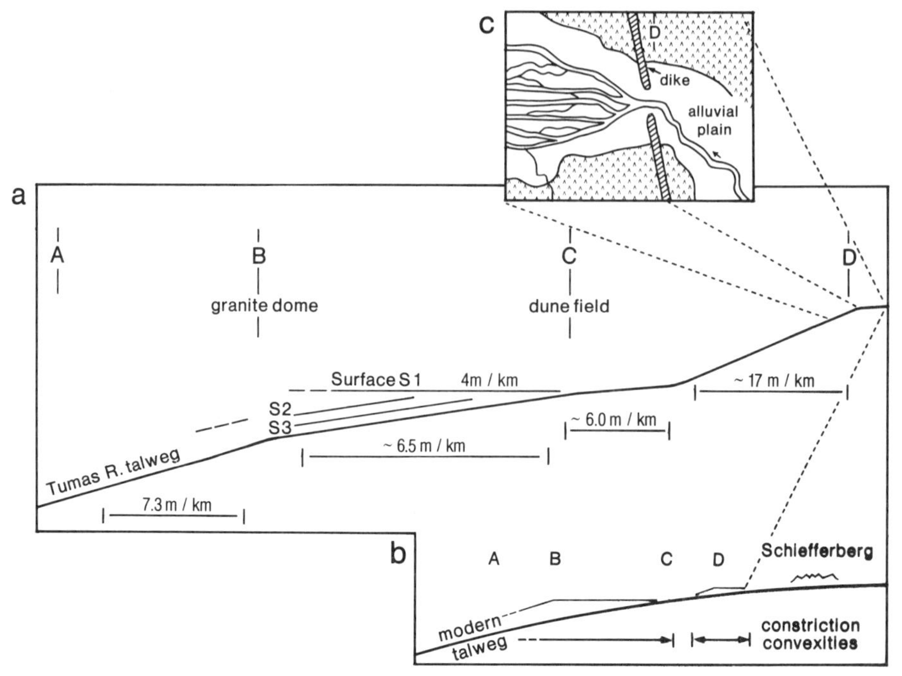

Fig. 19. Tumas River long profile illustrating convexity phenomenon. a. Slopes in each sector (A-D, fig. 4). b. Channel patterns at a convexity (D, fig. 4): straight/sinuous pattern upstream of dike constriction and distributary braided pattern below (left). c. Schematic profiles—localized convexities of lower Tumas basin superimposed on steepening general profile.

specifically in the sector B-C. Low terrace remnants, 1-3 m high, occur in the vicinity of point A (figure 4) and near Tumas Vlei.

Longitudinal and Transverse Profiles

Against the background of increasing downstream gradients of the Tumas (Stengel 1964; Goudie 1972)—a characteristic of many arid stream profiles (Mabbutt 1977; chapter 5)—the lower Tumas basin displays four distinct sets of long profiles: a steepest set in sector C-D upstream of the dune field where the stream is not incised (figures 4, 19a); a fossil set of very low declivity represented by the oldest terrace surface in the central, incised sector of the study area (sector B-C), and two intermediate sets represented by the modern channel of the incised sectors A-B and B-C. In the steepest sector (C-D, figure 19a), gradients decrease from highs of 17 m/km to 6 m/km at point C; stream width reaches extremes of 1000 m; and stream habit is notably braided. The high terrace in the central incised sector (B-C) displays the lowest gradients (4 m/km), whereas adjacent modern talwegs average 6.5 m/km with a contrasting nonbraided, straight-to-sinuous habit. The modern talweg and highest terrace diverge at an ill-defined knick-point in the vicinity of the small dune field (point C, figure 19a). Channel gradients are slightly higher downstream, averaging 7.3 m/km in the wide, unincised plain of sector A-B, although the main channel trace is not braided.

Transverse sections of the incised sector B-C display four surfaces. The above-mentioned high terrace is an alluvial surface that slopes downstream, whereas the lower three represent micropediments characterized by transverse gradients sloping toward stream beds. These four surfaces are referred to hereafter as S1, S2, S3, and S4 (figure 20a, b). S1-S3 are capped by recognizable gypsum crusts.

Surface S1 is preserved best in the center of the valley between the Tumas River and its main north tributary (figure 21a, b), and lies as much as 20 m above modern talwegs at altitudes of 325-285 m. It is almost completely removed by erosion in the western half of the central area, but notable outliers, of probable S1 affinity, can be seen at points across the north side of the valley plain up to 6 km west of the incised area (west and northwest of point B, figure 22).

The intermediate surface S2 spans much of the horizontal distance (up to 0.5 km) and vertical distance between S1 and modern talwegs, and terminates 2-5 m above talwegs, indicating that incision of approximately 10 m took place in the most incised areas in the vicinity of point B (figure 21b). Major outliers of this surface persist in the central and western sections of sector B-C (figures 21b; 23a, b). The main reason for the existence today of the S1 and S2 remnants is that they were heavily cemented by gypsum at the time they were current.

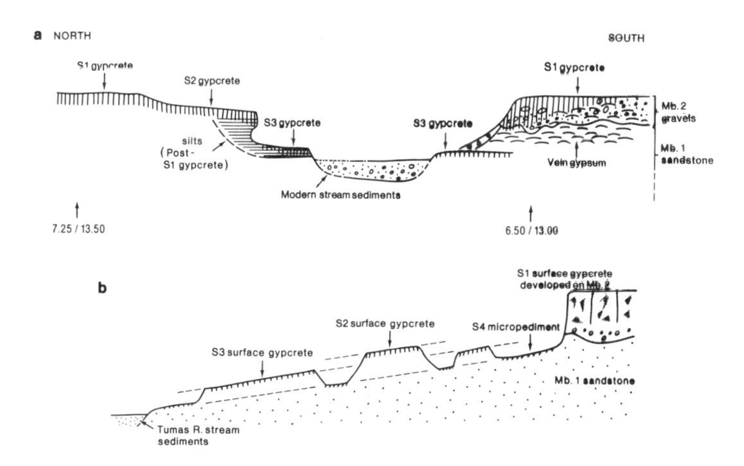

Fig. 20. Cross profiles. a. Composite along N-S lines 7.25 and 6.50. b. Schematic, illustrating three relict surfaces (S1-3) and current surface (S4).

Fig. 21. Geomorphic surfaces (S1-S4) of the central lower Tumas basin. a. S1 (arrows background), S3 (arrows foreground), S4 (foreground). b. S1-3 visible left, above stream-bed scrub; S2 outlier right; thin sheetlike Member 2 gravel underlies S1.

The most massive and resistant forms of gypsum crust develop preferentially within 100-300 m of drainage lines (see chapter 7). The remnants thus record prior positions of Tumas River channels and tributaries, and document the fact that major S1-related channels, with a WNW flow direction, lay 2-4 km north of the present Tumas channel (figure 22). Tributaries now flow southwest across the plain to reach the present southerly location of the main channel.

Surface S3 is areally smaller with shorter slope lengths than those of S2 (figure 22). It parallels the latter in slope, lying only 1-2 m below S2 (figures 20b; 24a, b).

Surfaces S3 and S4 each indicate modest incision of 1-3 m by the Tumas River. S4 is now more extensive than either S2 or S3, especially to the west (sectors A-B and some of B-C, figure 22) where upper surfaces have been destroyed; S4 often encircles S1 scarplets. S4 is situated from 2 m to less than 1 m below S3 levels (figure 24a, b). It is the current or submodern surface. Small alluvial fans form no part of the landscape of the lower Tumas basin (*contra* Wieneke and Rust 1976). Distributary flow on the valley plain produces thin veneers in the area of points A and C where some flows have transgressed the low but well-defined banks of the Tumas River.

River Metamorphosis and Arid Streams

Because the independent controls of stream behavior—bedrock confinement, discharge and sediment type—"can change significantly within a short distance" (Schumm 1981:157), stream morphology must be considered reach by reach rather than by drainage system (Schumm 1977). Of a comprehensive list of fourteen channel patterns proposed by Schumm (1981), "the change from one type of channel pattern to another should be relatively common" (Schumm 1981:157) within one watershed, as a result of the operation of independent variables.

The lower Tumas River course offers a prime example of stream metamorphosis, provided that certain assumptions are recognized. (1) Since discharges commonly decline (Leopold and Maddock 1953), and may cease altogether in arid zone streams (Mabbutt 1977), hyperarid systems often violate the assumption of downstream increase in discharge on which the relationships of stream morphology are based. This caveat appears not to invalidate the present analysis, because downstream attenuation of discharge is negligible in the comparatively short reaches of the Tumas stream under investigation. (2) Channel classifications are dependent mainly on studies of sandy river beds (Schumm 1977). Stream metamorphosis relationships nevertheless appear to hold in a general way, even in streams where caliber becomes coarser than sand for some reaches, as is the case in the Tumas River bed. (3) Finally, duricrust cappings can create indurated units that be-

a

Fig. 22. *a.* Granite dome (near point B, fig. 4) with well-developed S2 surface. Modern S4 surface foreground. *b.* S1 (a-a); S2 (c-c); S3 (d); S4 (e'); granite dome (b); Husab Mt. (f) (aerial photograph: U. Rust 1975, with permission).

b

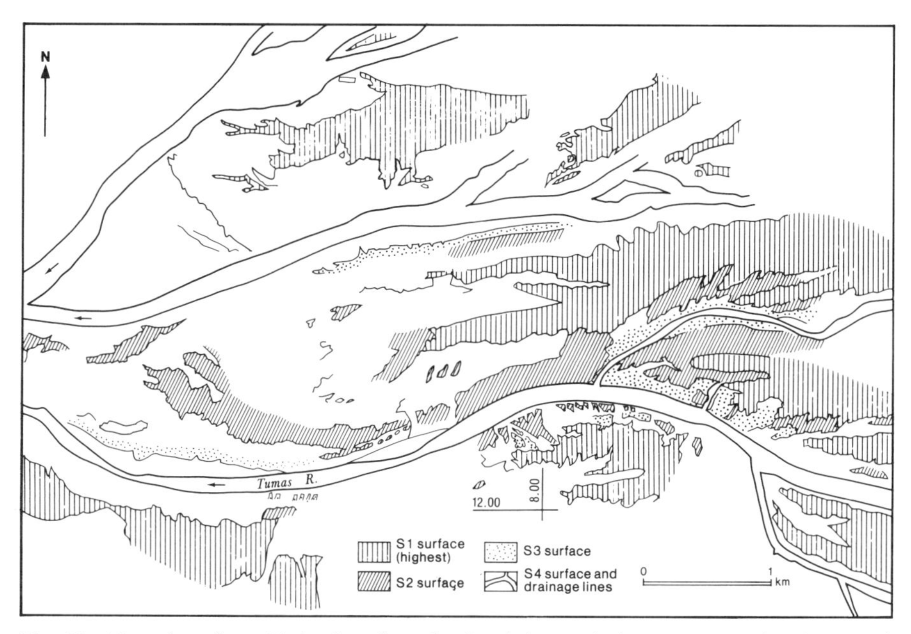

Fig. 23. Map of surfaces S1-4 where best developed in proximity to one another in central study area (sector B-C, fig. 4).

Fig. 24. Younger Surfaces S3 and S4 developed parallel with one another, separated by small vertical distances. a. S3 (duricrust) and S4 (modern) foreground, S1 (skyline).

have more as bedrock than alluvium (VanArsdale 1982). However, duricrusting of stream banks in the Namib Desert generally appears to be accommodated in that parameter of channel control termed bank cohesion.

The lower course of the Tumas River thus appears to be amenable to analysis in terms of accepted channel-metamorphosis norms, despite being an arid and endoreic catchment.

Interpretation of fluvial features in the lower Tumas basin must address two main questions. The first concerns the diachronic change from broadly aggradational styles to surface inactivity and linear degradation. The second concerns morphogenetic oscillations between active incision and morphostatic pedogenesis as the degradational phase runs its course.

Longitudinal Profiles

Aggradation

The Model of Constriction Convexities. Common to large and small streams, convexities in the long profile are often located at points where valleys are laterally constricted. Lower gradients occur on the upstream of such constrictions and steeper gradients downstream (figure 19c). Bryan (1925) commented on this characteristic of desert streams in the American Southwest. Flood plains in the Namib Desert are narrowed at points where resistant lithologies such as marble and quartzite ridges and dolerite dikes intersect valleys.

In the Namib Desert a further relationship exists: stream channel morphology tends to nonbraided habits on upstreams arms and braided habits on downstream arms of convexities (figure 19b). Localized, near-surface water is sometimes evident from patches of scrub in river beds at or immediately upstream of some constriction points.

In the model proposed here, aggradation in Namib Desert streams generates zones of deposition centered at points of valley constriction, zones that are morphologically convex in longitudinal profile. The cause of this phenomenon may lie in reduction of discharge velocities locally at valley constrictions. Deposition is then encouraged on the upstream side of valley constrictions.

The three sectors of the study area, A-B, B-C, and C-D, largely reflect elements of the model. The Tumas River is characterized by steep slopes to the east (C-D, figure 19a). Not only is sector C-D the downstream arm of a valley constriction at D (figure 19a), but field evidence shows that it is well supplied with notably coarse (cobble) debris shed from steep, flanking pediments that are slightly incised and surmounted by inselbergs. These controls, and especially that relating to coarse bed load, appear to be sufficient to explain the relative steepness of the C-D reach.

The B-C reach can be explained in similar terms. A major valley constriction at the granite dome (point B, figure 19a) appears to control the position of a long-profile convexity, low declivities of S1 occurring upstream of B and steeper declivities downstream. Modern slopes downstream of B are steeper than those of S1 and steeper than modern slopes between points B and C, suggesting that segments A-B and B-C conform on a large scale with the pattern of gradients of constriction convexities. It is suggested that the valley constriction at point B represents a larger, degraded convexity.

Sector B-C is interpreted as having undergone a long period of deposition, in the form of upper units of the Tumas Formation.

Depository Retreat. The depository represented by the main Tumas convexity—by its size arguably a major node of deposition—has become inactive and indeed degraded from its pristine form by subsequent stream incision. The depository appears to have retreated upstream a distance of at least 10-20 km to the vicinity of the dune field (point C, figure 19a), and further upstream.

The main evidence for present-day deposition in this area is a series of unconsolidated, unchannelized sediment gravity flows representing the distal termini of numerous discharge episodes that penetrated no further downstream than the dune field. The gravity flows, 100-200 m wide, are characterized by pebble and cobbly quartz lag surfaces that appear as high-reflectance tongues on aerial photographs. Their positive microrelief produces consistent downstream convexity of contour lines. Less viscous discharges are channeled between individual flows. Younger flows are lighter colored as a result of deflation and the surface concentration of quartz pebbles. Older flows, truncated by the younger, are darker as a result of lichen and patina growth (figure 25).

Consanguineous eolian deposits, in the form of the small dune field (figure 25), occur in this area of significant present-day fluvial sediment delivery.

Retreat of the locus of a major depository in an arid basin is generally taken to imply a reduction in discharge energy. In the situation of a narrow, arid-core desert, it is reasonable to suppose that climatic rather than other allogenic controls might have caused proximal retreat of a depository, because of the sensitivity of arid systems to fluctuations in moisture input. Geological literature is replete with examples of distal-proximal shifts of depository in ancient arid environments, and these are overwhelmingly interpreted as climatically induced unless good reasons exist to invoke other controls. Summarizing the long and intensive desert stream research undertaken at the Bardai Research Station in the central Sahara, Busche and Hagedorn (1980) have reiterated the view that seems to be well established, that geomorphic systems are especially vulnerable to change at the arid end

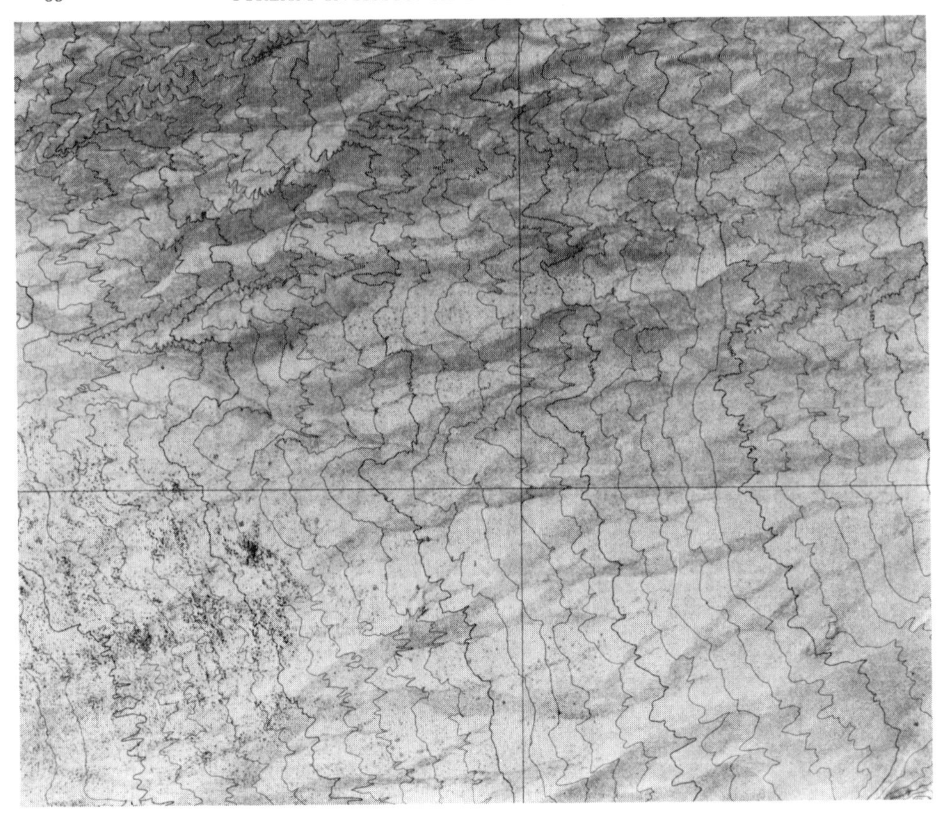

Fig. 25. Detailed aerial photograph showing several episodes of recent sediment gravity flows (right, top) upstream of small dune field (bottom left) (C, fig. 4). Scale: 2 km grid; 1 m contour interval (Anglo American Prospecting Services and Aircraft Operating Company, with permission).

of the precipitation continuum because so many geomorphic thresholds are transgressed.

Thus climates became drier. Climates somewhat wetter than those of today appear to have been responsible for laying down the Tumas Formation, in the form of sedimentation centered at the convexity at point B. Present hyperaridity forced recession of the depository inland as far as sector C-D, or further.

Incision

The phases of linear incision into, and soil development on, water-formed surfaces in the Tumas valley yield information on the degree of climatic change in the catchment. The energy required to initiate incision into sediments at the distal end of the Tumas basin could have involved tectonics and base level fall. Because of the existence of a rock bar across the mouth of the Tumas River immediately upstream of the end-point playa, and the localized nature of incision in the Tumas basin, the effect of these extrinsic controls seems less likely than that of variable discharge under the influence of climate. Internal dynamics of the system appear to have focused incision phases at constriction convexities, such that the convexity is progressively destroyed as a feature in the long profile of the Tumas stream. Incision of this localized type belongs to the class of fluvial behavior termed "autocyclic" by Beerbower (1964), and "episodic adjustments" by Schumm (1976, 1977, 1979, 1981).

Under conditions of erosion, the active fluvial zone is linear rather than areal, streams entrench alluvial flats, and constriction convexities become fossil forms with profiles inappropriate to the new fluvial regime. In this model it is suggested that rivers tend to remove local profile convexities (figure 19c) once the phase of active deposition ceases.

Explanations for localized incision into constriction convexities may relate to threshold stream gradients and discontinuous gully development. Bull (1979, 1988) has viewed river slopes in terms of threshold behavior and the tendency of rivers to achieve "critical gradient" profiles, defining critical gradient as the slope necessary for transporting the average amount of sediment supplied (Bull 1979). The concept differs from Knox's (1976) definition of a graded stream. In the latter, morphological constancy is the criterion of equilibrium (whether the stream is aggrading, incising, or neither), whereas the ratio of stream gradient to critical gradient is the key concept in the former (Bull 1979, 1988). Ratios greater than unity imply a tendency to incision, and below unity a tendency to aggradation.

Critical gradients are reduced progressively during deposition by a positive feedback relationship as streams deposit coarser fractions, thereby reducing the need for steeper gradients (Bull 1979). Such was the case during deposition of the Tumas Sandstone Formation. Present caliber of the load in the Tumas main channel is significantly coarser than Member 1 sands, being derived from Member 2 gravels, cobbles of detrital gypcrete and gravel from higher upstream. This difference in caliber probably accounts for the change from gentler slopes of the convexity to a steeper critical gradient during the incision phases.

Incision may be understood also in terms of the development of large discontinuous gullies. Schumm and Hadley (1957) have shown that incision is initiated in the steeper reaches of many small streams in the American West. There seems no reason to doubt that similar processes operate with respect to the flatter and steeper arms of constriction convexities of Namib Desert streams.

A further explanation probably relates to the phenomenon of incision downstream of a zone of sediment trapping. The new depository in the area of the dune field (point C) is a zone where gradient declines, distributary flow is pronounced, and the dune field obstructs discharge; these controls act in concert to trap sediment in the vicinity of point C. Such damming of natural waterways produces well-documented effects. One such effect is sediment starvation, which leads in turn to marked incision immediately downstream of barriers (Lane 1955; Makkavayev 1972). The influence of deposition at the dune field in the Tumas basin is deemed to be the same, with stream capacity slightly increased downstream and incision correspondingly enhanced.

The amount of vertical cutting declined at each successive stage of incision. The decline holds no necessary climatic implication of progressively reduced discharge; it is as much explained as a product of internal dynamics, with progressive relaxation of the tendency to incise as the convexity of the long profile was smoothed away.

The Tumas appears largely to have achieved "regrading" since S1 times, with both up- and downstream arms of the modern talweg convexity now displaying similar gradients. Low gradients of the upstream arm (B-C sector) have increased by 60%, from 4 m/km (S1) to 6.5 m/km (modern talweg). The downstream sector A-B is little different, but nevertheless steeper (7.3 m/km) than the B-C sector, a remnant perhaps of the downstream arm of the convexity. From this pattern, it might be predicted that future fluvial activity will continue the incision trend until the convexity of profile is entirely obliterated, provided the present degradational mode persists.

Channel Patterns

The interpretations concerning depository recession and stream incision are supported by examination of river channel patterns. The braided C-D reach gives way downstream to a nonbraided, straight-to-sinuous reach (B-C). Following conclusions in a recent review by Knighton (1984), such a succession is usually related to decreasing width/depth ratio, often associated with increased bank stability. In the case of the Tumas River, this trend corresponds not only to a five- to tenfold decrease in width parameter, but also to greatly increased bank cohesion by preferential gypsification of stream bank materials in reach B-C (see below, chapter 7). Furthermore, decreasing stream energy and slope are consonant with a change from braided to nonbraided habits in the downstream direction (Knighton 1984).

By contrast, gypsification of surfaces on the braided alluvial plain (sector C-D) is almost nonexistent, presumably as a result of the lateral migration of multiple, unstable channels cut in sandy and gravelly alluvium, and the apparent youthfulness of these incohesive deposits.

Decreasing stream load, especially decreasing traction load, is also associated with decreasing width parameter and is another control related to channel pattern change (Knighton 1984). A decrease in load is the likely consequence of topographic constraints that produce sediment starvation in the several square kilometers surrounding the dune field.

Schumm (1981) suggested that straight and sinuous channel types occupy the low-energy end of a continuum: they "reflect relatively low values of sediment transport, of bed-load to total-load ratio, and of stream power" (p. 19), whereas the braided type occupies the other end of the continuum and displays the opposite of each of these parameters. The two extremes of channel morphology juxtaposed in the Tumas River valley strongly suggest that active transport occurs in the C-D reach (figure 19a), and that bed load and stream energy are today much reduced downstream of this reach.

Whatever the merit of these models of intrinsic behavior, the extrinsic agency of channel evolution in hyperarid deserts is stream discharge. Since incision phases are predicated on sufficient stream discharge, it follows that phases of increased rainfall generated stream flow to points beyond the depository. Some idea of the degree of rainfall increase during incisional periods can be gleaned from the character of the petrogypsic crusts.

Transverse Profiles: Incision and Pedogenesis

Stream discharge capable of penetrating well beyond the dune field and thus incising the length of the major convexity centered at point B (sectors A-B and B-C) implies stronger flows than are presently characteristic of the Tumas River. Today, weak discharges follow the talweg trace without traversing the B-C sector. As in the Tsondab watershed (Stengel 1970), flow in the Tumas drainage appears never to reach its terminal playa with geomorphically effective flow.

Flows do not expand to fill the available width of the stream bed; nor have they done so during the period of time required for the evolution of a colluvial mantle on the banks of the main Tumas channel. Sufficient time has elapsed for gypsum and carnotite to have impregnated the mantle heavily. The existence of the mantle suggests that discharges have not covered the full width of the bed for at least some hundreds of years. Furthermore, the fragile gypsum crust capping the S3 surface appears unscathed up to the low banks of the present course of the Tumas River, suggesting that overland flow has been restricted sufficiently long for the development of a true crust.

Geomorphically insignificant discharge is undoubtedly characteristic of the present regime. Stronger discharges seem required to remove large quantities of sediment and incise the bed. Since three periods of incision are evident, it is plausible to suggest that discharge in the Tumas River has increased on three occasions. Incision stages are separated by periods of system decline and geomorphic quiescence with duricrust formation.

Evolution of micropediments S2, S3, and S4 requires comment. Their transverse orientation, the similarity of their slopes one to another, and the evenness of the surfaces suggest sculpture by surface runoff. Since runoff of the kind necessary for cutting such pediments is inimical to the preservation of young and fragile crusts, and quickly infiltrates less indurated crusts, it must be concluded that the micropediments represent environments different enough to prevent crust initiation, or, more likely, for incipient crusts to be removed by active slope processes. A reasonable assumption, which is consistent with the evidence adduced thus far, is that runoff was sufficiently vigorous to perform beveling.

In short, existence of the pediments adds to the impression that moister climates indeed occurred in the Tumas catchment—climates that generated through-flowing discharges strong enough both to permit major incision and to sculpt small pediments in relatively soft sediments.

Conclusion

Paleoenvironments

I have suggested that terracelike features and convexity of the longitudinal profile of the Tumas River in the vicinity of point B accord with a phenomenon of desert stream morphology, namely the constriction convexity. It is suggested that the convexity at point B is a localized feature representing a major depository of the Tumas basin, in a comparatively vigorous river system. The fact that the convexity comprises a series of mass flows, phenomena related to river end points, indicates, however, that the end point probably lay between the coast and point B.

Subsequent recession of the depository away from the coast is suggestive of decreasing stream energy and thus very likely of increasing aridity. The retreat distance may have been short (as little as 13 km), or the end point may have retreated tens of kilometers.

Duricrusting of the convexity surface and linear incision into it are phenomena best explained by recession of the basin depository, with a change to incisional modes of behavior in the lower Tumas stream. Concomitant gypsification of the S1 surface argues strongly for aridity.

Thereafter, alternating phases of incision and gypsic soil development suggest fully arid conditions with fluctuations to the moist side. There is no firm evidence of high water tables once incision had begun.

A picture of morphogenetic phases appropriate to climatic variation in the Central Namib begins to emerge. The nature of the arid phases probably parallels that of the present arid phase. Gypsum crust development—the subject of chapter 7—is the dominant process, with mobilization of both gypsum and near-surface carnotite by frequent fog. Eolian effects are lesser and comprise mainly the building of coppice dunes downwind (i.e., on the south side) of major sandy stream beds of the central Tumas basin. The development of reg surfaces, scarplets, and wind streaks, with minor deflation hollows, are related features. Water tables are apparently low, and stream flow is almost nonexistent in the study area. When discharge occurs, it is declining and accomplishes minimal work.

Moister phases in the study area, by comparison with arid morphostatic phases, appear to have been morphodynamic. They were characterized by stream discharges capable of incising and removing indurated sediments, and by micropedimentation along major stream courses. Stronger discharge was most likely generated by increased precipitation in the Tumas River headwaters, but some local increase is adduced from the development of the pedimented surfaces flanking the main stream beds.

In the Ubib drainage 140 km southeast of the study area, Hüser (1976) has likewise ascribed incision phases to periods moister than those required for soil development (see below, chapter 9). Mabbutt (1952:360) reached the same conclusion in the Ugab River valley 200 km to the north, referring the wettest incision phases to "semi-arid" precipitation conditions.

Fluvial Sedimentation in Arid, Confined Alluvial Plains: The Tumas Model

Attributes of the Tumas depository identified in this and the last chapter can be combined in a depositional model. With fine-grained lacustrine sediments almost entirely lacking, dominant constituents are fluvial sands and gravels, crudely bedded or structureless, with a conspicuous lack of channel forms. One member of the Tumas Formation was interpreted as the product of numerous gravity flows, unconfined by channels; the other unit as partly gravity flow and partly stream-flood in origin. As is common in desert streams, discharge decreased in the downstream direction as a result of infiltration, evaporation and the loss of tributaries. Increasingly viscous mixtures resulted, finally ceasing to flow. In Miall's (1985) architectural scheme, these are SG: gravity flow elements.

By avulsing laterally gravity flows accumulated to form a sheet—termed SH: sheet forms—across the entire width of the valley. With time, such sheets stacked vertically to form the deposit as preserved today.

Discharge specifically failed to reach a terminal playa at the low point of the basin and thus failed to generate lacustrine sediments. The reach of the valley floor in which deposition occurred is determined by the ratio between rate of discharge into the depositional basin and the rate of infiltration and evaporation. The reach where flow units were stacked is thus largely a function of climate. Climatic changes are manifested by migration of the depository reach in a distal direction during wetter phases, and proximally during drier phases—as is evident from many Namib streams.

Within the climatically determined depository reach, aggradation appears to have been focused preferentially at points where valley walls constricted the flood plain laterally. Longitudinal profiles of streams are convex to the sky at these narrowed points, termed "constriction convexities" in this study.

The rudimentary distinction between channel and overbank deposits can be lost as channels—by definition concave to the sky—decline in importance to the point of extinction. Indeed, an opposite trend seems to exist: convexity of bounding surface becomes a common if not overriding feature. At smaller scales gravity-flow styles of sedimentation yield convex-up units, and these in great number yield sheetlike lobes of sediment on the floor of an aggrading flood plain. Friend (1978) has given examples of convex lobes in ancient rocks. At larger scales, convexities of long profile characterize sedimentation in a confined valley setting, particularly at points of valley constriction. It is apparent that in settings not confined by valley walls, the convexity phenomenon will not arise.

The degree of valley confinement appears to influence facies frequency (Friend 1978) and facies type (Sneh 1983). This aspect of geomorphic setting is a growing theme in reconstructions of sedimentary environments (see, for example, Rust and Gibling 1990). Soil distribution, stream morphology, water table topography, vegetation, and uranium distributions were all influenced by location with respect to constriction convexities in the case of the Tumas basin.

Valley-bottom gravity flow deposits are said to be poorly preserved because subsequent river action obliterates them. In arid basins where such river action may be limited owing to lack of through-flow discharge to the ocean, there is no reason why gravity flows should not attain preservation in the geologic record.

Soils developed on valley-bottom surfaces when active aggradation ceased. Multiphase, indurated duricrusts characterize arid environments so that pedogenic facies can become important constituents of fluvial sequences, as Allen (1974) has shown for alluvial plains of the Lower Old Red Sandstones. Under wetter climates and an erosional regime, stream flow once again penetrated beyond, and incised into, the duricrusted aggradation convexities, reestablishing flatter long profiles.

Various lines of evidence support the interpretation that sedimentation in the Tumas basin took place under arid conditions. In the absence of an alluvial fan setting, gravity flows signify the zone in a valley where discharge finally ceases to flow. Declining channel depth, to the point that channels disappear, has been argued to be a feature of declining fluvial systems (e.g., Friend 1978; Nichols 1987) including arid basins. Reworking of dune sand—as apparently occurred in the Tumas Formation—supports the notion of an eolian-dominated environment, although such reworking need not typify all settings of Tumas type.

Apart from an arid climate, the major extrinsic control of sedimentation is a tectonic setting of shield rather than classical basin-and-range type in which great distances typically separate watershed hill-country from a terminal playa.

Blatt et al. (1980:632) have observed that "it seems probable that there is also a type [of depositional environment] characterized by ephemeral streams in semi-arid climates, intermediate between humid floodplains and typical semi-arid alluvial fans . . . , but the model for this type is not yet well developed." In fact, several models, including the Tumas model above, have begun to fill this gap in fluvial sedimentology. Some models concern radial drainage like that on alluvial fans, others concern linear patterns. Some incorporate overt climatic control from humid to arid settings. Yet others compare confinement by valley walls with unconfined settings.

Fluvial depositional environments in the gap identified by Blatt et al. (1980) include terminal fans (Mukerji 1976; Friend 1978; Graham 1983; Parkash et al. 1983; Olsen 1987), wet fans (e.g., Galloway and Hobday 1983), large inland plains termed deltoids (Whitehouse 1944), "immense alluvial cones" (Iriondo 1984; Baker 1986:300; Whitehead et al. 1990; Wilkinson 1991), braidplains and braided rivers (e.g., Miall 1977, 1978, 1987; Rust 1978; Rust and Koster 1984), sandflats (Hubert and Hyde 1982; Smoot 1983), and piedmont alluvial plains (Williams 1970; Rachocki 1981). Sneh (1983:99) described sediments in the "terminal floodplain" of the great, central wadi of the arid Sinai peninsula. This flood plain is, however, unconfined, with sediments quite unlike those in the Tumas Formation.

Recent discussion of fan deltas has specifically sought to clarify the multifarious similarities and differences between classic alluvial fans and braided rivers (McPherson et al. 1987, 1988; Nemec and Steel 1988). Plains dominated by anastomosing streams in both arid and humid climates have been identified as depositional environments unique in their own ways, differences in stream density apparently determined by climate (Mabbutt 1977; Rust 1981; Rust and Legun 1983).

River basins in which discharges do not reach the ocean, nor even, as in the case of the Tumas sediments, the playa at the topographic low point of the basin, are not specifically incorporated in any of the settings mentioned above. The Tumas model differs from the rest in the particular combination of arid climate, sedimentation on a flood plain, distance from a mountain front, and a linear setting confined by valley walls, the walls in this case comprised of flanking pediments.

Because their scales differ so widely, many of the above-mentioned depositional models cannot be directly compared, but they serve to illustrate the variety of models that have been constructed and also to point to some of the unexpected controls identified by field studies. The variety and nature of fluvial settings begins to appear a decade after Baker (1978) and Miall (1980) voiced concern about the narrow base of environments that underpin fluvial sedimentology. But this work remains in its infancy—"fluvial sedimentologists have simply not studied enough rivers to be aware of depositional styles in the full range of tectonic and climatic conditions" (Miall 1987:4).

Chapter 7

DESERT SOILS AND URANIUM MINERALIZATION

The parched eviscerate soil
Gapes at the vanity of toil.

—T. S. Eliot, *Four Quartets*

The supergene accumulation of gypsum, uranium, and iron compounds in the upper levels of the Tumas Sandstone Formation each allow deduction of certain minimum environmental conditions demanded for their formation, although none of these minerals is very well understood in their natural occurrence. Conclusions can be drawn concerning precipitation conditions and high, falling, and fluctuating water tables. Furthermore, a relative chronology of soil-forming events is available from a sequence of gypsum crusts.

The three mineral suites are both pedogenic and nonpedogenic in character. The gypsums and phenomena related to iron compounds are conveniently described in terms of master soil horizons, but the uranium concentrations are not, being usually invisible macroscopically and highly discontinuous both vertically and laterally. Uranium mineralization has been described mainly in terms of assay values, uranium equivalents by scintillometer and by particle-track counts captured on buried, sensitized film. The uranium distributions are less illuminating with respect to past environments than either the Fe or $CaSO_4$ compounds.

Three idealized polypedons illustrate typical occurrence and horizonation of the three groups of minerals as they occur in the parent lithologies. These polypedons form the basis for subsequent detailed description and discussion of each mineral. The polypedons describe typical profile development in three geomorphic settings, (1) on interfluves, (2) in stream bank localities and (3) in wide, possibly fossil stream beds in the central study area. A notable feature of these profiles is that they are hosted only by sediments of the Namib Group. Surrounding older rocks are often devoid of a regolith cover susceptible to soil formation, a control of soil distribution by substrate found in other warm deserts (Cooke and Warren 1973). All the gypsum crusts represent an absolute accumulation of gypsum in different crystalline form.

All three polypedons are dominated by the specific morphology and development of gypsum crusts that can be described in terms of Watson's (1979, 1983, 1985) three categories, all of which are found in the Tumas valley:

1. Crusts composed of mesocrystalline crystals of lensoid habit developed on hillsides; within this category are included fibrous, prismatic, powdery and alabastrine types.
2. Desert rose crusts *(croûte de nappe)*, comprised of large interlocking crystals in the southern Namib (Kaiser 1926), displaying small, desert rose clusters in the Tumas area; these are associated most plausibly with evaporation at a water table; powdery and alabastrine types are related.
3. Horizontally bedded crusts of probable lacustrine origin, displaying graded bedding, the larger mesocrystalline lensoid crystals replaced upward by microcrystalline and alabastrine types; this type is very minor in the study area.

All types display Munsell color hues, from white in the case of pure fibrous types, to pale yellow in the case of powdery types, to gray and light brown in older, weathered massive alabasters.

Idealized Profiles

Where it exists, the macrostructure of all three polypedons takes the form of anticlinal features or domes of low amplitude known as "pressure structures." Usually less than 1 m across, domes are often connected, imparting a wavy cross-sectional pattern of domes and basins (figure 26). Vertical pressure-induced joints are a less prominent characteristic of the macrostructure, usually confined to stream-bank and -bed polypedons.

Following the Food and Agriculture Organization (1977), Horta (1980), and the Soil Conservation Service (Guthrie and Witty 1982), the gypsic horizons are designated "Y" in figure 26.

Hillslope Polypedons (figure 26a)

On valley-side slopes and interfluvial surfaces of the alluvial plain, capping crusts are typically of mesocrystalline type. Crusts vary from incipient to cemented, the latter Petrogypsic Gypsiorthid (Soil Survey Staff 1975) being an accumulation up to 1 m thick. In pit walls this type appears as a honeycomb, the honeycombs sometimes formed into larger pressure struc-

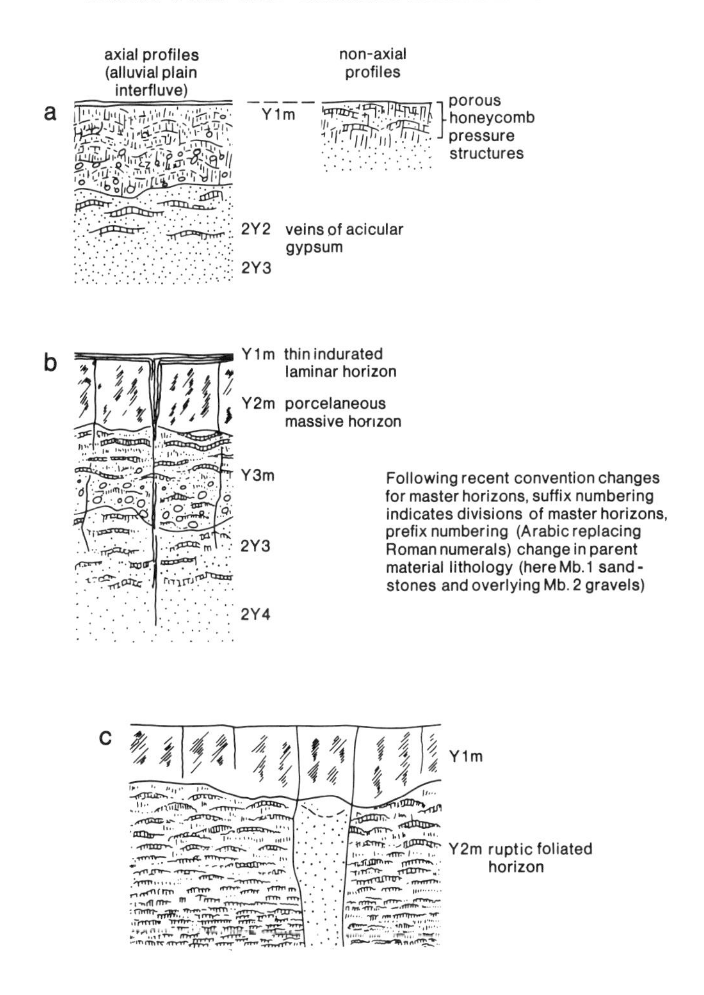

Fig. 26. Idealized gypsum polypedons. a. Hillslope. b. Stream bank. c. Stream bed.

tures. It contains variable amounts of impurity made up of Members 1 and 2 host materials; it often acts as host to the uranium mineral carnotite.

On interfluve flats of the alluvial plain it is overlain on inactive surfaces by a nongypsiferous, vesicular layer 1-3 cm thick. It is underlain by the strongly reddened horizons of the Member 1 sandstone, itself often a host for carnotite.

Stream Bank Polypedons (figure 26b)

Indurated, porcelaneous, alabastrine crusts up to 1.5 m thick are exposed along all larger, incised stream courses in the study area. They may contain gypsum contributed from active discharge and base flow. These crusts are nonporous, vertically jointed, and show an intricate relationship with powdery gypsums lower in the profile. Concentrations of gypsum can be very high (up to 90%) and these accumulations are undoubtedly displacive in character. Intense veining by mesocrystalline types also takes place in the lower parts of the gypsum horizon. The underlying red sandstone displays a thick gleyed horizon, in some cases with numerous mottles. Except for the uppermost massive horizon, in which radioactivity is characteristically below background levels, carnotite can occur throughout the profile.

A restricted variant of this form is the softer surface crust, exposed at gypsum diggings known as von Stryk's claims (point C, figure 4). Here a pure surface crust 50-70 cm thick is apparently undergoing exposure and disaggregation into corestones, in a process described by Watson (1979, 1985). Immediately beneath, pure, powdery, structureless gypsum, usually less than 1.0 m thick, has been mined. (Locally in the same area, an impure powdery gypsum overlies the crust.)

Stream Bed Polypedons (figure 26c)

Host materials in the wide, braided stream beds at the east end of the study area are heavily impregnated with gypsum. In the type-site pit (2.875/12.350), the upper 1.0-1.5 m consists of high concentrations of gypsum in both mesocrystalline and alabastrine forms. Within the underlying red sandstone is developed an unusually thick (2.5 m) network of gypsum veins (up to 1 cm thick) made up entirely of small pressure structures with wavelengths up 6-8 cm. Horta (1980) has termed these "foliated" structures. In the study area the structures decrease in length and amplitude with depth, the lenses becoming thinner and the network denser. Interstitial sand lenses are cemented by cryptocrystalline gypsum, and carnotite is prominent throughout the profile.

A dominant feature of this profile is its ruptic character: gypsiferous red sand wedges, 40-80 cm wide, penetrate through the entire depth of the profile and lack gypsum-related structures (figures 26c, 35). The wedges are taken to be the fills of desiccation fissures of major proportions. The red sandstone host exhibits high uranium values.

Gypsum Crusts

Many researchers have mentioned the gypsums of the Central Namib, although detailed studies are few. Indeed the gypcretes received no treatment in a world survey of duricrusts (Goudie 1973), perhaps because of their comparatively small areal distribution. During the last two decades, the association of some types of supergene uranium mineralization with calcrete and gypcrete has established the latter as a desert soil with economic and theoretical interest in its own right. In particular, attention centers presently on the topics of sulfate origin (never more so than in the case of the Namibian gypsum crusts: Martin 1963; Besler 1972; Cagle 1975; Watson 1979, 1983, 1985; Wilkinson 1981), refinement of duricrust description, and depositional environments responsible for the generation of crust types (e.g., Bellair 1954; Coque 1962; Tucker 1978; Cody 1979; Watson 1979, 1983, 1985).

Watson (1983, 1985) has suggested that the Namib gypsum crusts are distributed along the coast up to 50 km inland, beyond which surficial calcretes are dominant. Only halite efflorescences are more soluble, and these are widespread in washes from the coast 30 km inland (Watson 1985). Calcium sulfate concentrations as high as 90% are common. Six of the purest samples of twenty from the study area averaged 79% $CaSO_4$; van Wyck (1969) reported large volumes of relatively pure gypsum, some samples with 92-93% purity in and around the study area. Fifty-three samples from the central Namib gave maximums up to 90% pure gypsum, recognizable crusts with concentrations as low as 20% (Watson 1983).

Purity decreases from the surface downward, estimated values declining from 90% in the upper 30 cm to 45% at 1.5-2.0 m depth (van Wyck 1969). Six samples from Member 1 red sandstones at 4 m depth gave sulfate concentrations of 3% and less, however. Scholz (1968) showed that crusts contain a range of carbonate inclusions with an average in the Outer Namib of 12% (Scholz 1963), the purest forms with less than 10% (Watson 1983).

Soils of the Central Namib have been classed most recently as Gypsic Yermosols by the Food and Agriculture Organization (1977), and somewhat misleadingly, as solonchaks and calcareous sands and loams (Harmse 1978). The hillslope and stream bank polypedons accord with the *Soil Taxonomy* definition of Petrogypsic Gypsiorthids.

Although Besler (1972) and Watson (1983, 1985) have followed Martin (1963) in proposing that airborne marine salts are the source of the Namib gypsums, bedrock sources may well be important (Wilkinson 1981), quite apart from the presence of constituent minerals from which gypsum can be formed: Nash (1972) and Cagle (1975) reported primary anhydrite in ancient rocks exposed near surface; Gevers and van der Westhuyzen (1931) reported raised sebkha deposits of halite and gypsum on the Swakop canyon rim; Reuning (1925) has described small natural sulphur deposits at Richthofen and Birkenfels near Goanikontes on the lower Swakop River; and Cagle (1975) and Schaller (1987) have identified a dominance of nonmarine sulfur isotopic ratios in the surface gypsums of the study area.

Gypsum Crusts and Topographic Surfaces

A toposequence of pedogenic crusts is a marked characteristic of the Tumas alluvial plain gypsums. Successively younger geomorphic surfaces

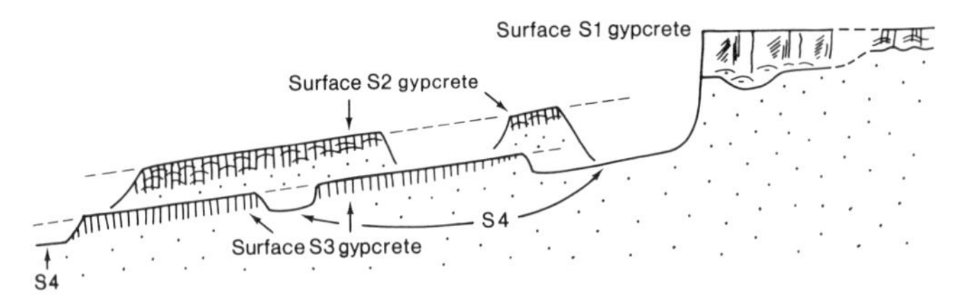

Fig. 27. Surfaces S2 and S3 show increasingly thick crust development with distance downslope. Thinner horizons upslope are more easily removed erosionally.

(S1-S4) display different crust types (figures 21b, 23a, 27). Beyond the alluvial plain, such topographic differentiation does not exist, and the widespread gypsum crust, where it has developed on the younger sediments, must be polycyclic, as are the older crusts in the study area. Whereas Namib calcretes first develop as subsurface horizons (Blümel 1982), pedogenic gypsum crusts appear to develop from the surface downward (*contra* Watson 1983, 1985), concentrations decreasing with depth.

Surface S4. Incipient crusts are developing on the youngest geomorphic surfaces of the alluvial plain (S4). Small mesocrystalline, and occasional well-developed fibrous types, as yet not segregated into clusters, develop on and within newly exposed red sandstone to a depth of less than 10 cm. Gypsum concentrations are low and the surface is rendered soft, porous, and "puffy" (Goudie 1972). Color of the parent material is little changed (figure 28).

Surface S3. The lowest and least prominent of the fossil surfaces (S3) hosts the least well developed crust, whereas S4 accumulations do not qualify as crusts in terms of the 15% minimum content defined by d'Hoore (1965). The S3 crust is 10-20 cm thick (figure 24), with numerous inclusions of the underlying parent material; it is less puffy, more indurated, and more porous than the incipient crust, with colors of higher value and lower chroma, a trend that continues as the content of gypsum increases. Well-developed mesocrystalline veins have formed by strongly displacive growth. The latter is a well-attested process (Watts 1978; Watson 1985; Machel 1985). Incipient pressure structures are evident.

Surface S2. Surface S2 is armored by a crust 50-60 cm thick with well-developed pressure features; this too is highly porous, but purer and more resistant than the prior crust type (figures 21b, 29); it contains a greater variety of crystal types and shows simple catenal development in the form of thickening downslope (figure 27).

Surface S1. The highest surface (S1) lies as much as 20 m above the larger talwegs and supports a crust 1.5 m thick, comprised of upper and lower horizons. The crust has a coherence that has preserved the surface as the most prominent younger feature in the lower Tumas valley (figures 13,

Fig. 28. Early stages of large crystal growth on youngest surfaces (S4) cause incoherence of surface materials: see gilgai surface (foreground) and indented vehicle tracks. Induration capable of supporting vehicle weight characterizes older surfaces.

Fig. 29. S3 crust meets S1 crust (near B, fig. 4). Incoherent S4 crust middle ground.

21a, b). Two major crust types occur in catenal relationship on surface S1, namely stream bank and hillside profiles.

The stream bank occurrence is the purest and most dense; its upper two-thirds is composed of microcrystalline gypsum and displays a massive, alabastrine, plugged horizon. In many places the massive horizon is overlain by a thin (up to 3 cm) laminated upper horizon. The laminations extend vertically downward along the largest joints. Joints are usually less than 1 m deep, however, and give the crust a columnar appearance (figure 30a, b, c). Some joints reach a depth of 2.5-3.0 m, penetrating underlying lithologies. Apart from the effects of such joints, the massive crust is also the least porous.

Displacive growth is the dominant mode of accumulation, as is evident in that host structure and even pressure structures are entirely obliterated, clasts are brecciated, and gypsum concentrations gain very high values (figure 30a, b, c). Volumetric changes related to simple gypsum buildup and to mineral phase changes in the gypsum-bassanite-anhydrite sequence also account for jointing and pseudo-bedding features (figure 13a).

Beneath the massive, upper horizon, which can exceed 1 m in thickness, often separated by a pseudo-bedding discontinuity, lies a less intensely gypsified horizon, composed of a dense network of veins of fibrous gypsum (figure 31a). Volumetric changes are thought to account for the development of microcrystalline, powdery types (Watson 1983), and indeed these are found in the study area beneath the massive petrogypsic horizons. Pressure structures are often evident within the mass (figure 31b).

Hillslope-interfluve profiles, set well away from streams, are also highly indurated, but unlike the massive type, are highly porous in conformity with their typical mesocrystalline composition. These profiles are less pure and display little jointing and no laminations, but abundant pressure structures (figure 32a, b). The porosity of this type, maintained by the displacive growth of relatively large crystals, appears to be related to higher uranium values than the nonporous type; the mesocrystalline type most often displays visible yellow coatings of carnotite (figure 32b).

Catenal relationships. Crusts on footslopes of younger surfaces are thicker and somewhat more resistant than those on upper parts (figure 27). Such slopes seldom reach more than 500 m in length in the incised sector of the study area (B-C, figures 4, 22). Surface S1, however, shows stronger differentiation. Thicknesses of the massive alabastrine crusts diminish to extinction within a distance of 100-300 m from stream banks. They are replaced by the indurated honeycomb variety (figure 33).

Reasons for the phenomenon of optimal crust development along stream banks appear to be, first, the existence of more gravelly Gms facies of Member 2 along drainage lines, which are conducive to heavier concentra-

Fig. 30. 1-2 m-thick S1 crust (stream bank polypedon). Massive porcelaneous and jointed in upper parts (a, b, c) with pressure structures generated by numerous veins of fibrous gypsum in lower parts (a). Absolute increase in gypsum volume causes disaggregation of host gravels (c).

a

b

Fig. 31. Lower horizon of streambank pedon comprises dense network of fiber gypsum veins. a. Incipient pressure structures—redder. b. Denser network with some pressure structures (arrows)—pinker.

a

b

Fig. 32. Highly porous honeycomb-like crusts underlie interfluvial S1 surfaces. These are mainly lensoid mesocrystalline crusts. a. Thinner pressure structures. b. Thicker pressure structures, with rare visible coating of ubiquitous yellow mineral carnotite.

Fig. 33. Gradation of S1 crust types: massive porcelaneous crust (left side of pit) thins to corestones with distance from drainage line—wedges over porous honeycomb type (right side of pit). Member 1 red sand dump in background.

tions of sulfate. Second, gypsum is concentrated nearer drainage lines in the course of time, even on minimal slopes, by overland flow. This mode of surface accumulation has been documented in the specific case of gypsum crusts (Risacher 1978). The increased concentration of sodium salts along drainage lines has undoubtedly also aided the process of gypsum infiltration and consequent development of the massive alabastrine type, since the presence of halites increases the solubility of gypsum fourfold (Zen 1965).

Stream discharge also concentrates gypsum along riverbeds and banks, both detritally and in solution: occasional large clasts of massive gypsum crust occur in stream bed loads and in unconfined flows on upper surface S1 in the vicinity of point C (figure 4); solutional concentrations are apparent where microcrystalline gypsum has produced extreme and compact (i.e., nonporous) cementation of some stream bank walls.

A final and probable cause of the catenal relationship is the greatly enhanced wetting generated by inclined and especially vertical surfaces during fog events. Present stream banks are notably steep and consequently moist well into the day in south-facing localities (such evaporatively cooled overhangs provide much-used, and thus enlarged, shelters for oryx and jackal). Wetting of this kind, many nights per year, probably results in added mobility of gypsum, to the point of promoting complete infilling of interstitices. It is shown below that moisture causes active lateral and vertical translocation of soluble gypsum. Thin (5-15 mm), secondary, crustlike efflorescences of mixed halite and gypsum coat many of the steeper stream-bank walls. More important, wetting of this type, rather than proximity to water tables as suggested by Watson (1983), appears to be the more plausible explanation for development of small, pebble-sized desert rose crystals that occur within the honeycomb gypsum crusts.

Parallels with Caliche Toposequences. The trend from minimal gypsum accumulation on the youngest surface to massive crust development on the oldest surface parallels the main elements of caliche K-fabric stages documented by Gile, his coworkers (Gile et al. 1965, 1966, 1981), and Machette (1985). Accumulations are thicker on land surfaces of increasing age in and around the Rio Grande valley of southern New Mexico. The progression described from the Tumas valley probably relates to two dominant controls: (1) the progressive supply of gypsum to a particular point by various processes, and (2) the length of time the crust has been exposed to wetting. These appear to lead to cementation and plugging, via clast coatings, veining, and cluster formation. Plugging and thin, laminated horizons are both features that closely parallel the final and most advanced expression of the carbonate K-fabrics.

Infrequent precipitation, in the form of rain rather than fog, probably leads to lateral movement of water on the upper surface of the plugged horizon. This hypothesis satisfactorily explains the laminated horizon. Gile et al. (1965, 1966) have suggested an origin by similar means for the K21m carbonate horizon.

Not only does the gypcrete sequence parallel trends that are documented in calcrete profiles, but parent-material caliber exerts a control similar to that in the case of caliche. Gravels and gravelly sands in the study area exhibit the most massive, tabular crust development, whereas sandy and silty facies of Member 1 exhibit thinner, less massive crusts with a greater proportion of thickly developed fibrous gypsum veins in pressure-structure form.

An approximate, correlative sequence in the Namib gypcretes of the carbonate duricrust stages presented by Gile et al. (1965, 1981) is the following: thin, nonindurated, fibrous types of high porosity —> indurated, mainly porous pressure structures of medium thickness and various mesocrystalline types —> thick, porous, indurated, honeycomb, mesocrystalline types on sandy substrates; or, on coarser substrates, crusts with thick, nonporous, jointed, massive upper horizons and mesocrystalline, veined lower horizons.

The sequence may proceed further, since powdery gypsum probably results from repeated solutional attack on larger crystals with rapid reprecipitation (Watson 1983, 1985) in microcrystalline form.

Paleoenvironmental Implications

Several conclusions can be drawn from the existence of the crusts and their types. Whereas other duricrust minerals have a more enigmatic environmental meaning (Watson 1985), there has long been consensus that the existence of gypsum crusts can be taken to indicate aridity (Jenny 1929; Coque 1962; Cooke and Warren 1973; Watson 1979, 1983, 1985, among others). Watson (1983:155) has stated that "gypsum crusts are indicators of very specific climatic environments generally where annual rainfall is less than 200 mm/yr. and where there is a monthly excess of evaporation over precipitation throughout the year." He noted that gypsum crusts are not in equilibrium with present southern Tunisian climates of 150 mm annual rainfall, under which they are suffering solutional attack.

Present mobility of gypsum. Sulfate crusts on the youngest surfaces appear to be pristine and in the process of formation under present conditions. Powdery crusts demonstrably undergo dilation through wetting by fog, as evidenced by gilgai relief in unconsolidated, 30-cm-thick surficial gypsum near the small dunefield (point C, figure 4). Excavation showed that gilgai ridges are centered over joints in the underlying massive crust. Goudie (1972) mentioned similar features related to large desiccation polygons near Gobabeb. Detrital cobbles and boulders of massive gypsum are truncated by karstic attack where exposed subaerially.

The most persuasive example of present gypsum accumulation, however, is that of a colluvial mantle exposed in trench 3.50/11.50 on the inner bank of a bend in the main southern arm of the Tumas River (figure 34). Here a 0.75-m-thick mantle is heavily impregnated with veins and clusters of acicular gypsum, some fibers reaching 3 cm in length. These masses are postdepositional growths: since the mantle postdates the final cutting episode of the stream, it and the secondary accumulations can be considered relatively very recent.

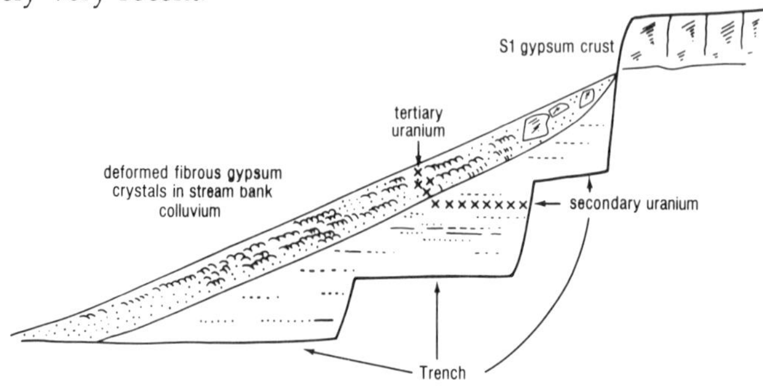

Fig. 34. Deformed gypsum fiber clusters in stream-bank colluvial mantle. Bank of Tumas main arm, central study area.

The sulfate source is undoubtedly detrital, derived from the massive gypcrete scarp upslope. The agent of solution is apparently fog precipitated on the bank slope, the steepness of which promotes heavy fog precipitation. Tumas river discharge has never cut the foot of the mantle. En masse downslope deformation of gypsum fiber clusters indicates movement of the mantle by gravity during crystal growth (figure 34), movement possibly enhanced by the invasive growths.

Not only is gypsum mobile under present conditions. In the same trench, a vein of carnotite in undisturbed Member 1 sandstone has acted as source to a daughter vein that extends macroscopically up through the entire thickness of the colluvial mantle (figure 34).

There seems little doubt that present conditions are conducive to crust growth. Surface wetting undoubtedly promotes growth of gypsum crystals at the surface. Deeper veins are preserved due to the lack of flushing. It can be assumed that wind- and water-borne detrital gypsum, and fog- and groundwater-borne salts have provided the upper meters of the Tumas Formation with ready supplies of gypsum. Scholz (1963), Martin (1963), Besler (1972), Rust and Wieneke (1976), Rust (1979, 1980), and Watson (1983) have also ascribed crust formation to the present arid climates. Goudie (1972) implied the same.

Chronosequences. The chronosequence of four ages of gypsum crust, on different-aged surfaces, provides evidence for geomorphically quiescent periods (*sensu* Erhart 1967; Butzer 1976b) disrupted by periods of stream incision and micropedimentation. The remarkable inactivity of the Tumas stream since the final phase of incision is apparent from the existence of a complex colluvial mantle developed on a major stream bank. Apparently, gypsum is mobile today, to the degree that it moves readily on subaerial surfaces, collecting downslope and disrupting exposures of red sandstone. Despite the mobility of gypsum, present rainfall provides too little flushing of soils to remove gypsum from the surficial 1-3 m.

As a result, surfaces of different ages support crusts of markedly different thickness and internal structure. Progressive accumulation strongly indicates that intervening phases of greater moisture were insufficiently long or intense to remove crusts solutionally.

The important distinction between nonflushing, gypsum-dominated soil-moisture regimes nearer the coast versus soils characterized by a greater degree of flushing in the Inner Namib (calcrete) echoes precisely a distinction established early this century between extreme arid desert cores and their less arid peripheries (Busche and Hagedorn 1980).

In light of these observations on the different ages of gypsum crusts, Besler's (1972) view that the crusts predate canyon incision probably does not apply to the youngest and may not apply to the others.

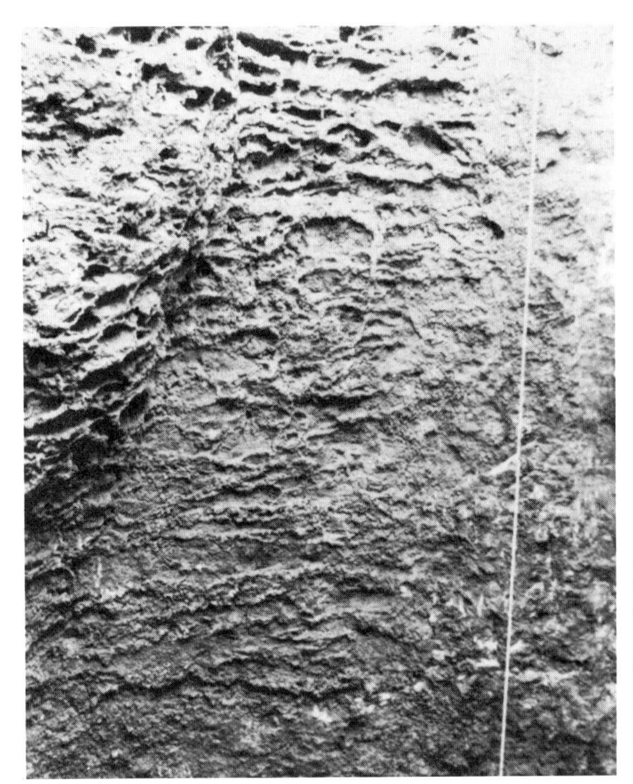

Fig. 35. Foliated structures in stream-bed pedon developed best near surface. Vertical ruptic break of foliated structures shown right (left of marker twine). Scale: pen top center.

Foliated Horizons. A final issue concerns the very deep foliated, root cast-rich, stream-bed profile (figures 26c, 35). Reported as recent deposits from Algeria (Horta 1980), such foliated horizons probably relate to progressively falling water tables (Horta 1981). Overburden weight is probably responsible for limiting pressure structure amplitude and thickness, and causing an increase in vein number with depth.

It has been noted that the foliated horizon is ruptic, that is, penetrated by deep (3.5 m), vertical wedges (40-80 cm wide) of sand lacking foliated structure. It seems likely that the period of falling water tables documented by the foliated gypsum was followed by deep desiccation fissuring. The fissures were filled by slope wash and fluvial sands, materials that show the typical red sandstone transition zone (80 cm thick) from red to gray hues (see below). This zone indicates the existence of a subsequent, permanent, near-surface water table. Whereas abundant root casts in the foliated profile indicate valley bottom (shrubby?) vegetation, the lack of root casts or foliation in the wedge fills suggests water tables significantly higher than they are at present, but below root level.

Goudie (1972) has documented a series of very large patterned ground polygons (averaging 8-9 m in diameter, some reaching 20 m) on the 42 m terrace at Gobabeb "associated with calcareo-gypsiferous crusts . . . many feet deep" (p. 21). The polygons display up to 40 cm of relief and a visible pattern at surface, probably the result of burial and subsequent widespread gypsum crust development. Goudie (1972) ascribed the existence of the polygons to greater moisture in the past. A mass of evidence on the origin of large polygons supports the desiccation model (Christiansen 1963; Cooke and Warren 1972; Goudie 1973; Neal and Motts 1967).

Red Desert Soils

Strong reddening and an underlying reduced zone are the dominant pedogenic features of the Member 1 red sandstone. One of these characteristics appears to hold promise for paleoenvironmental interpretation. The red sandstone becomes less coherent with depth and is best displayed in pits located on the higher-lying alluvial plain interfluves above the reduced zone. Macroscopically, there is no trace of horizonation through 4 m (and even 7 m in one locale) of red sand (see Munsell color descriptions in chapter 5; figures 13b, 31a, 33) in the deepest interfluve pits.

Its consistently sandy nature is a feature suggestive of Ferallic Arenosols in the Food and Agriculture Organization (1977) terminology, and Psamments (possibly Typic Xeropsamments) in the *Soil Taxonomy* (Soil Survey Staff 1975). Upper horizons, if they existed, were probably removed everywhere when the Member 2 depositional episode commenced. That pedogenesis of Member 1 had already begun is suggested by gleying of the red sandstone and lack of reddening or gleying of the superjacent Member 2 gravels.

Microscopic examination of the red sandstone shows two phases of reddening. The first is manifested as a discontinuous coating of iron oxide on quartz grains, and the second by weakly crystalline intergranular material composed of an orange-red mineral closely bound up with an off-white, amorphous pasty substance, possibly gypsum, carbonate, or silica compounds. The latter material yielded no usable X-ray diffraction results (by Gandolfi camera scan), but the former suggested a rare, unnamed iron hydrogen phosphate hydrate mineral, $FeH_3(PO_4)_2.4H_20$ (Lehr et al. 1966).

The presence of the mineral is significant because it indicates pedogenic reddening. The consistently low clay fractions in the red sand, however, prevent classification of this deep horizon as either oxic or cambic, and make appropriate its designation as a sandy, horizonless Entisol with incipient clay buildup. Red coloration does not necessarily indicate significant iron oxide accumulation, since 0.4% by weight is sufficient to impart such hues (Walker and Honea 1969). The color is a reflection of the alteration of copious ferromagnesian grains derived during deposition from nearby source terranes.

The minimal requirements for pedogenic reddening are (1) a supply of iron and (2) texture appropriate for oxidation (Pye 1983). In the case of this Psamment, iron-bearing minerals are abundant and texture is conducive to aeration and positive Eh potentials. The grain coatings may also represent in situ weathering and slow dehydration of amorphous or poorly crystalline ferrihydrite mother compounds (Gardner and Pye 1981; Pye 1983). Even if these grains are detrital, they may well have reddened in or around the central Tumas basin.

The presence of nondehydrated amorphous compounds suggests that present soil moisture conditions are responsible for their formation. If this is indeed the case, caution is required in making any inferences concerning past environments based merely on the presence of the red bed. Oxide release may continue slowly in horizons deep enough to be unaffected, in a nonflushing environment, by gypsum accumulation.

In reviews of an extensive literature on red beds, Gardner and Pye (1981) and Pye (1983) concluded that iron oxide release can occur under both arid and humid tropical conditions, though release is probably slower in the former. Low clay percentages in the red sand suggest very slow rates of soil formation, although soil-forming processes do operate today. The redder eastern dunes of the dune Namib have attracted attention (Logan 1960; Barnard 1973; Besler 1977, 1980, 1984; Lancaster and Ollier 1982) as possibly older and immobile dunes. Pye (1983) noted that immobility is a requirement for reddening, and Besler (1972) noted that Namib dunes are presently moist at depth. Minimum requirements are thus met for reddening under present aridity.

These considerations pertain equally to the buried red soil documented by Scholz (1972) 5 km northeast of Gobabeb, and by Koch (in Scholz 1972) on the Walvis Bay–Rooibank road, both of which Scholz (1972) has interpreted as possible indicators of past, moister soil regimes. There is little doubt that periods of higher groundwater level and increased throughflow would have enhanced the reddening process, but the red profile alone does not necessarily indicate greater moisture in the Tumas basin.

The existence of a thick, diffuse zone of reduction, possibly a polygenetic horizon, is implied by the ubiquitous, 10-15-m-thick zone of paler Munsell values (figures 13b, 36). This zone underlies the red horizon and is best exposed in riverbanks along the most deeply incised arms of the Tumas River. Iron compounds that have undergone reduction display characteristically gray, blue, and green hues (Needham 1978). Over vertical distances of tens of centimeters, reds of the red sand give way to pinks, which in turn give way in places to some or all of light greens, grays, and light mauve sub-

Fig. 36. S1 crust characterized by corestones and amorphous gypsum is well developed in Member 2 gravels, giving way with depth to red sands of Member 2. The latter are pink at depth where proximity to past water table levels has encouraged gypsum emplacement and reduction.

zones. Few to many distinct, coarse mottles occur. The boundaries of this zone are diffuse: the transition between reds and most intensely reduced colors varies from 1 to 2 m.

For these reasons, as well as consideration of environments conducive to the precipitation of uranium and vanadium (below), it seems reasonable to suggest that this zone represents a hydromorphic zone. Indeed, colors accord with those defined for the G: gley subsoil horizon in *Soil Classification* (MacVicar et al. 1977) definitions. The zone does not achieve the standards required by the *Soil Taxonomy* (Soil Survey Staff 1975) for an Aquent, but this probably is explained by subsequent vestigial oxidation since the reducing environment, or environments, disappeared.

Considering that the reduced zone lies a full 10-20 m above the present water table, it seems to provide prime evidence for the existence of a higher water table, or water tables, at some time in the past.

Evidence for rising water tables is encountered in pit 6.75/12.417, where 1 m of structureless, reddened sand is underlain, at 2.5-3.2 m depth, by a rectilinear network of joints that separate corestones 3-10 cm in length. The corestones are pink and the network joints lined by light blue and green hues. The network is underlain by a structureless, light gray-green horizon of the red sand. This sequence apparently documents a rising water table that reduced the red sand entirely below 3.2 m, but only partially along the joint network, leaving the corestones with hues reminiscent of the upper red horizon.

Uranium Mineralization

Uranium mineralization occurs almost exclusively within the upper 20 m of the surface (Wilkinson 1980; Hambleton-Jones 1984), in the zone of groundwater fluctuation between present water tables and the surface. Support for the existence of high water tables in the past comes from the distribution and equilibrium status of uranium minerals to which the Tumas Sandstone is host. The dominant uranium mineral in the Namib supergene province is carnotite $[K_2(VO_4)_2(UO_2)_2.3H_2O]$, which occurs as yellow coatings on sand grains and gypsum crystals (figures 32b, 37). In a geomorphic categorization of world supergene uranium depositional environments, Toens and Hambleton-Jones (1984) have classed the Tumas occurrence with the fluviatile group (in the confined valley subgroup), rather than with lacustrine and pedogenic types.

Within the fluvial setting, the Langer Heinrich (Gawib Flats, northeast of the study area) and Tumas occurrences seem to be located upstream of major valley constrictions (Wilkinson 1977; Hambleton-Jones 1984). High positive disequilibrium ratios (greater chemical assay values than expected from uranium spectrometric equivalents) characterize barrier zones and indicate comparatively recent deposition (Hambleton-Jones 1984).

Not only does a similar valley constriction occur in the Tumas case, but the uranium is also in a state of disequilibrium at Langer Heinrich (Hambleton-Jones 1984) and in the Tumas units (Levinson 1985). From the similarity of their topographic settings, it seems reasonable to suppose that the same processes, especially those related to high water tables and soil suction, are responsible for the Tumas uranium bodies. The existence of the generally tabular body within 20 m of the surface alone is suggestive of higher water tables (figure 20a). Reducing conditions per se beneath a water table are known to precipitate uranium (Boyle 1984). Furthermore, reduction-oxidation fluctuations, conducive to adsorption of complexing species (Mann and Deutscher 1978), are best promoted at water tables.

Carbon dioxide partial pressures are reduced at those points where groundwater is forced nearer the surface; such pressure reduction is known to lead to dissociation of mobile uranyl ions, with consequent carnotite precipitation; dissociation is also enhanced by negative Eh potential, and by changes in pH (Mann and Deutscher 1978; Boyle 1984; Briot 1984; Carlisle 1984; Hambleton-Jones 1984).

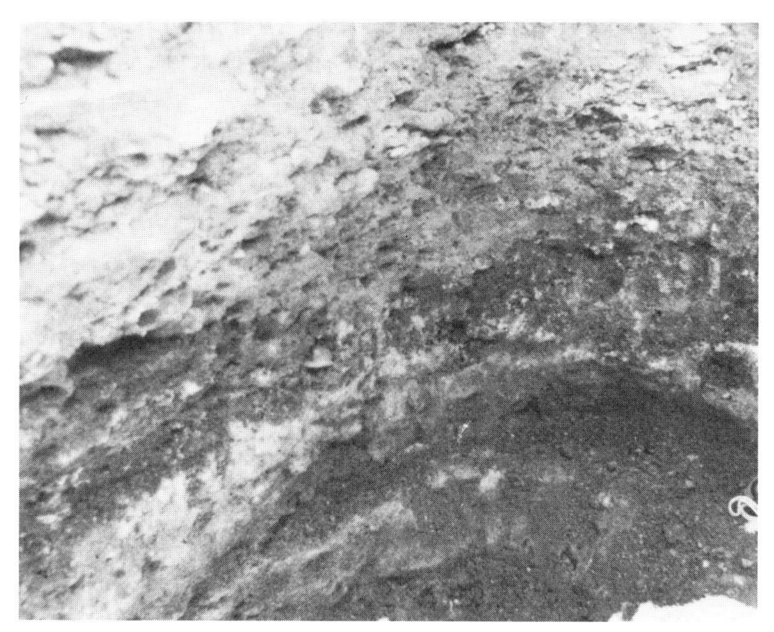

Fig. 37. Diffuse stratiform zones of carnotite coat upper gypcreted levels of Member 2 gravels.

Another conducive mechanism is that of evaporation of groundwater (Hambleton-Jones 1984). Since more intense evaporation occurs at near-surface water tables, such localities in the lower Tumas may be expected to display preferred mineralization. Adsorption of complexing ions such as vanadium is strongly promoted in the presence of organic material (Coker and DiLabio 1979; Boyle 1984), and organic material in a desert fluvial environment is likely to be many times more abundant in riverbed locales where water levels rise to within the root zone.

In the case of the Langer Heinrich deposit, Hambleton-Jones (1984) concluded that mineral precipitation resulted from changes in reduction-oxidation conditions, from adsorption of vanadium and from evaporation of groundwater. The last was judged the dominant mechanism in small, near-surface mineral bodies.

In sum, unlike gypsum soils, aridity per se is not indicated by the presence of uranium minerals in the Tumas sediments, since supergene uranium enrichment occurs in all climates (Boyle 1984). However, climate does determine the style of the deposit, in this case a fluviatile calcrete/gypcrete style (Boyle 1984). The acknowledged controls of above-background uranium mineralization are indeed suggestive of higher water tables in the study area at a time, or times, in the past.

Further, the preservation of these mobile minerals is evidence for nonflushing environments since deposition. Because water tables are conducive to the operation of so many contributary processes, the 20 m thickness of the richest carnotite zone is suggestive of water table fluctuations, probably within much smaller amplitudes, but with numerous superimposed episodes. Since water tables are presently relatively deep-seated, they are probably less effective as precipitating environments.

The thick carnotite zone thus provides only generalized, undatable evidence for high water tables in the past. Two 400,000 B.P. dates, calculated on samples in the upper 2 m of the red sandstone, are meaningless because of isotope disequilibria (Levinson 1985). But they indicate, at least, that uranium has been mobile roughly within the last half million years.

Conclusion

The change in chemical environment from calcium carbonate in the Leeukop Conglomerate to calcium sulfate in the overlying Tumas Sandstone may represent modern syngenetic effects of precipitates in arid depositories. It appears to be more likely, however, because of the probable difference in age of the two Formations, that the carbonate environment belongs to an earlier period in the history of the Namib when such soils were widespread.

Less equivocally, it can be concluded that Namib climates have been sufficiently arid for gypsum accumulation—as against calcrete accumulation—for the length of time required to build up the petrogypsic, topographically prominent S1 crust. That the oldest crust, morphologically quite distinct from the youngest crust, is extant in the landscape indicates that arid, formative climates have not been disrupted, or have been disrupted little by intervening wetter events.

Rainfall of 200 mm/yr. for an extended period seems absolutely excluded, since such a regime would flush all gypsum from surface host units. Gypsum is completely replaced by calcrete under the present climatic regime of 150 mm/yr. in the Inner Namib. The laminated gypsum horizon, designated Y1m, which extends down major fissures in petrogypsic horizons, is probably the product of flowing water and is best explained as the result of occasional rainfall.

It has been argued that the existence of reddened sandstone in the study area and further south does not imply wetter or hotter environments in the Central Namib. Occasional rainfall and groundwater flow explain the wetting of the sandstone; at deeper levels full desiccation may take many seasons to accomplish, and indeed may not occur, so that reddening could easily result. Furthermore, surface duricrusting processes, processes explicitly confined to nonflushing soils, do not reach depths of >2.5-3.0 m.

The juxtaposition of red substrates with gypsum-whitened surface horizons, commented upon by Scholz (1972), Koch (in Scholz 1972), and

Sandelowsky (1977), probably relates more to geomorphic changes that have reexposed those buried, reddened sands by erosion. Reddening thus seems more a function of sufficient depth of burial beneath duricrusted, surficial layers, as well as the necessary supply of moisture, ferromagnesian minerals, and appropriate host texture.

Visible above modern talwegs, the thick, diffuse, and often mottled zone of bluer hues in the Member 1 sandstone, is a zone that has been interpreted as evidence of one or more high, permanent water tables. By contrast with overlying red hues, this horizon provides strong evidence for the existence of high water tables, probably before the final three incision phases of the lower Tumas basin, and probably before the first phase. The thickness of both the zone and its diffuse upper boundary suggests fluctuating horizon development, or multiphase ascent of water tables to these general levels.

Pits in eastern parts of the study area illustrate the complexity of the chronology. Most gypsum crust occurrences in the study area are pedogenic, but the foliated, groundwater-related crust with pedotubules seems to provide strong evidence for high, but falling, water tables in a section of the stream where water tables are now 10 m below surface. The foliated profile has been deeply ruptured by apparent desiccation fissures. These phenomena indicate the presence of dissolved gypsum in near-surface water. The degree of increased flow implied by a high water table must be reconciled with the fact that rainfall cannot have been so high as to cause flushing of surface gypsums.

The profile thus suggests a trend from a period of slightly moister-than-present climate, with high water tables (perhaps semiarid in the headwaters along the Great Escarpment), to falling water tables (successive foliated horizons), to one of hyperaridity characterized by low water tables, desiccation fissuring and sandy fissure fills. Fissure fills display evidence of reducing environments in lower levels, and since they lack pedotubules, indicate a subsequent rise in water table, to levels too low to promote abundant shrub growth. The massive surface gypsum that caps the foliated profile can probably be correlated with S1-surface petrogypsic crusts. This suggests (1) that the foliated profile, and probably the reduced horizon, predate the S1 surface, and (2) that water tables were high during the preincision phases of the lower Tumas valley.

The uranium-rich zone of Member 1 is generally coincident with the near-surface reduced zone, likewise diffuse, and likewise indicative—with somewhat less certainty though great likelihood—of water tables higher than present. It is probable that the uranium zone represents several episodes of high water table, some or all coeval with those represented by the reduced zone. Some uranium remobilization in groundwater has occurred within the last half million years.

Because of strong rainfall gradients from east to west across the Namib, high water tables can indeed be reconciled with low rainfall in the lower Tumas basin. Present precipitation in the Inner Namib (none fog-derived) is 150 mm (Lancaster et al. 1984), nonfog totals falling to less than half this figure in the study area only 80 km to the west. If rainfall increased uniformly, or if gradients steepened, high groundwater levels—generated by increased rainfall along the Great Escarpment—are not incompatible with gypsum soil development, or at least with the maintainance of these crusts in the Outer Namib.

An interesting sidelight on regolith mobility (induced by gypsum encroachment) concerns the development of convexo-concave slopes on sediments of the Namib Group. Up to 50 cm of surface material is disaggregated to the degree that a colluvial mantle is generated, which is sufficiently mobile to give rise to smoother slopes. Extensive slopes on the north side of the Tumas valley in the A-B sector are underlain by relatively noncoherent Namib Group lithologies prone to gypsum buildup. These slopes display comparatively rare, nonrectilinear slope profiles. Similar mantles and convexo-concave slopes have been observed in the Atacama Desert of northern Chile.

Chapter 8

YOUTHFUL EOLIAN FEATURES OF THE TUMAS FLATS: IMPLICATIONS FOR PAST AND PRESENT WIND REGIMES

The [east] wind was so strong that we could lean against it. A red veil of sand rose from a small sand dune into the blue sky. To the south a sandstorm was raging and the dunes were covered with a reddish mist.

—Henno Martin, *The Sheltering Desert*

In dem . . . Swakoptal . . . scheint der Ostwind besonders heftig zu sein. Er hat hier in der Talaue viele 1-2 m höhe Dünen um die Busche herum angeweht. . . . 1917 war der Staubsturm einmal so stark, dass man nicht über die Strasse gehen konnte.

—F. Jaeger, *Geographische Landschaften*

Youthful geomorphic features of the study area include a congerie of fluvial features (pebble gravel spreads on interfluves and minimal talweg discharge) and eolian phenomena (reg surfaces, wind streaks, small dunes, and sand veneers). All of these are consistent with modern morphogenesis in the Tumas basin and they provide a background against which to view past environments. One set of features in particular, however, has provided information on past eolian environments in the Central Namib. Analyses of wind streak and dune patterns help to provide chronological perspectives on phases of dune formation and are, accordingly, the subject of this chapter.

Evidence of wind activity in the dune-free Tumas basin and elsewhere on the northern, eastern, and southern flanks of the Dune Namib, has been thought of as nonexistent or minor in importance. In fact, such evidence contributes to an understanding of the age of eolian activity and wind regimes in the Tumas basin and in the dune sea, issues of intrinsic interest for study-area paleoenvironments. Further, this kind of evidence contributes to a solution of the problem of formative wind environments of dune patterns within the sand sea, arguably a major event, or series of events, in the reconstruction of past west coast environments.

The analysis concerns (1) alignments of eolian features in the study area, and (2) alignments of particular sets of dunes in the complex of dunes that constitutes the Namib sand sea. The orientation of linear dunes in particular has been employed widely in recent years as a palimpsest of formative wind patterns, especially in the reconstruction of past wind regimes. From the arrangement of three different but coherent sets of linear dunes in central southern Africa, Mallick et al. (1981) and Lancaster (1981b) have proposed that the subcontinental anticyclone has occupied three different positions during the late Cenozoic.

Because of the potential importance of linear dune alignments in such reconstructions, two opposing theories of linear dune development in the Namib form the basis of discussion in this chapter, in light of perspectives derived from wind streak phenomena in the Tumas basin.

The Dune Namib stretches 340 km along the Namib coast between the port towns of Lüderitz and Walvis Bay and extends 100-120 km inland, covering an area of 34,000 km^2 (Lancaster 1983). The northern boundary is fairly abrupt at the Kuiseb River (figure 2). Three broad groups of dune patterns have been recognized in the northern half of the dune sea (Barnard 1973; Besler 1980, 1984; Lancaster 1983): (1) a major central tract dominated by large, north-south-aligned linear dunes (50-150 m average height, spaced 1.2-2.8 km apart; Lancaster 1982a) of specific interest to this study, (2) a transverse set in the 30-60 km-wide coastal tract, and (3) an eastern tract characterized by star dunes (figure 2). Several dune types have been identified within these groups (Besler 1980, 1984; McKee 1982; Lancaster 1983), as have many patterns of regularly recurring dune types and orientations (Besler 1980, 1984; Lancaster 1982a, 1983).

The opposing theories of linear dune formation in the central tract may be termed the synchronic theory of Lancaster, and the diachronic theory of Besler. In a series of studies, Lancaster (1980, 1981a, 1982a, 1982b, 1983, 1985) has applied modern wind regime data to existing dunes and dune patterns, clearly implying that the dunes are equilibrium forms related to present circulation of the atmosphere. If the dunes are current forms, then linear dune alignments hold little interest for reconstructing past wind flow patterns. The opposite holds in the case of Besler's diachronic reconstruction.

Lancaster has promoted the view, via a "general model" (Lancaster 1982a; 1983:284) of linear dune development, that the major north-south linear dunes of the Namib are (1) fashioned under bimodal wind regimes and (2) aligned obliquely to stronger SSW winds. McKee (1982) supported the interpretation of a bimodal origin. The SSW winds have been extensively

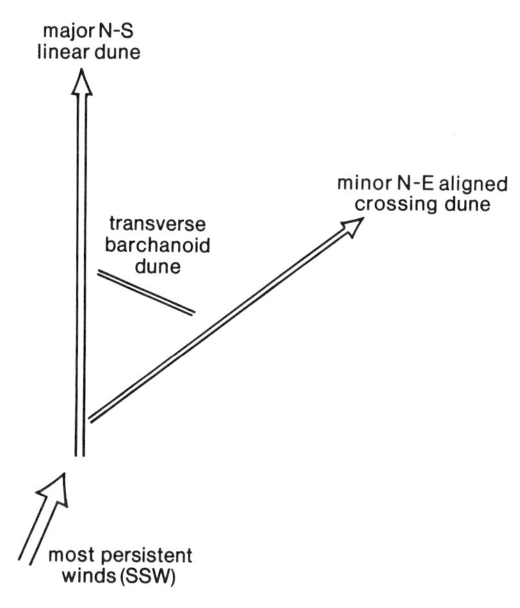

Fig. 38. Lancaster's model of dune alignment with respect to formative SSW winds: linear dunes (large NS-oriented and smaller NE-oriented dunes) aligned obliquely, linguoid ridges in the angle aligned transversely. Arrows show directions of propagation (after Lancaster 1983).

documented in the dune field as the dominant sand-moving winds (Breed et al. 1979; Lancaster 1983, 1985; Lancaster et al. 1984; Ward and von Brunn 1985). The theory of dune propagation at angles oblique (20-30°) to resultant winds, in a bimodal regime, has been derived from processual studies by Tsoar (1978) (see also reviews in McKee 1982 and Lancaster 1982a). In Lancaster's (1982a, 1983) model, three sets of dunes are ascribed to the action of the SSW wind, namely (1) the above-mentioned large, north-south linear dunes, (2) the minor linear dunes, aligned northeast-southwest ("corridor crossing dunes") and (3) small, transverse dunes in the angle between the linear sets (figure 38).

Lancaster's model coheres with the conclusions of a worldwide study (Breed et al. 1979) of the development of dunes under present-day wind regimes. Since wind speeds decrease markedly inland of the coastal tract (Royal Navy and South African Air Force 1944) and wind variability increases concomitantly (Breed et al. 1979; Lancaster 1985), the oblique propagation model has the theoretical support of representing moderate wind energies of the linear dune tract as a weaker, inland extension of the high energy, southerly quadrant winds of the coastal tract. In terms of world patterns, high velocity, unidirectional winds correlate well with areas of trans-

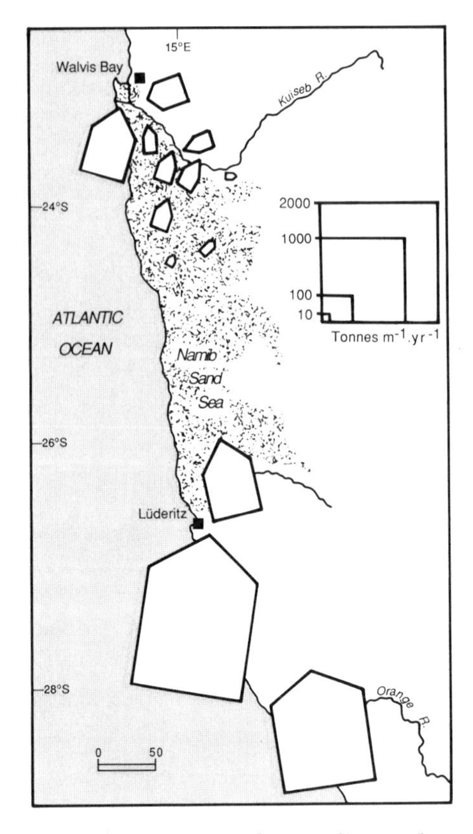

Fig. 39. Sand movement resultants in the Central Namib (after Lancaster 1985).

verse dune types, whereas linear dunes seem to be associated with intermediate energy regimes (Breed et al. 1979; Fryberger 1979).

Further, in conformity with present theory, the linear dunes are seen as products of a bimodal wind regime, which in the case of the Namib comprise the dominant SSW wind and the subdominant east wind, giving a resultant to the north (Breed et al. 1979; Lancaster 1979, 1982a, 1983, 1985; McKee 1982).

Indeed, only two sets of winds appear to monopolize the movement of sand in the Namib. These are the southerly and ENE-quadrant winds (Lancaster 1982a, 1982b, 1985) (figure 39).

Besler's (1976, 1977, 1980, 1984) theory relies on Taylor-Görtler boundary layer modeling, by which linear dune spacing in the Dune Namib demands higher-than-present wind velocities to generate horizontal helical vortices of appropriate diameter. Such vortices are regarded as formative for the chains of large linear dunes. Besler (1976, 1977, 1980, 1984) has proposed that higher wind speeds are best explained as part of more vigorous, Pleistocene atmospheric circulation. The existence of ventifacts in interdune corridors (known locally as "streets") and on the rocky plains is regarded as evidence of long-continued aridity (Besler 1976).

The two theories are also opposed in terms of probable sand mobilities as well: Besler (1977:52) regarded only dune crest sands as mobile under present conditions: "Im innern Erg bei Gobabeb (wandern) nur die Kammsande der grossen Dünen und bilden kleine Vorläuferdünen . . . während der Grundriss der Dünenzüge unverändert bleibt."

Besler (1977:52) wrote specifically of the "weitgehende Stabilität der unteren Dünenpartien." Besler (1977, 1984) also regarded the sand as locally derived (from both the underlying Tsondab Sandstone and continental shelf), and relatively immobile in the long term, on the grounds that five discrete granulometric "sand provinces" (Besler 1984:449) in the dune sea remain substantially unmixed.

This view receives support from the dating of hearth charcoal at an archaeological site on a dune 2 km west of Gobabeb (Sandelowsky 1977). The 12,800 B.P. date led Vogel and Visser (1981:55) to remark on the "unexpected stability" of the dune surface on which the site is situated. In his publications on wind and sand flow, Lancaster (1980, 1981a, 1982a, 1982b, 1983, 1985) has implied that the Namib dunes are forms in equilibrium with modern wind regimes, and has criticized Besler's diachronic theory as implausible (Lancaster 1982a, 1983).

Eolian Features on and around the Tumas Flats

Although the plains north of the Kuiseb River are strikingly free of dunes by comparison with the Dune Namib, they display several eolian features that are apparently forming under the influence of present-day winds. Except for the coastal dune cordon, which illustrates the continued effect of southerly winds, the northeast Berg Wind is the formative wind inland on the plains. Wind and sand resultants show the influence of this wind (Breed et al. 1979; Harmse 1982; Lancaster 1983, 1985), as do such features as small coppice dunes (tied dunes or *nebkha*), thin and discontinuous sand veneers on windward mountain slopes, offshore dust plumes, deflation hollows, ventifacts, wind-faceted boulders, and wind streaks.

That these features all show the imprint of the Berg Wind is clear: coppice dune tails extend southwest of obstacles and sand veneers blanket windward eastern faces of Rössing and Khan Mountains (Goudie 1972), the Chuos Mountains, and the Marmor Pforte Ridge (Wilkinson 1976). The sand blankets expand to the size of large dunes on the eastern slopes of some mountains in the Inner Namib (e.g., Saagberg; see Hüser 1976). These dunes are now fixed by vegetation, duricrusted by caliche, and fluvially incised. Sand-blast faceting is heavily dominated by northeast winds (Selby 1977b; Sweeting and Lancaster 1982). Several deflation hollows of significant dimensions (up to tens of square kilometers in area) occur near Rooibank and show preferred northeast-southwest orientations.

Of particular significance to the present argument, however, are wind streaks, features defined by Greeley and Iversen (1985:209) as "patterns of contrasting albedo [that form] as a result of various aeolian processes." The streaks are straight, narrow features of little or no relief, which show marked parallelism with one another. Wind streaks are usually associated with topographic obstacles to wind flow, are strongly elongated in the direction of the formative wind, and may result from grain size differences, grain mineral (color) composition, or eolian bedforms (Greeley and Iversen 1985). Analysis of streak orientation has provided basic data in the analysis of planetary atmospheric circulations, by which, for example, Thomas and Veverka (1979) have demonstrated the asymmetry of Martian atmospheric hemispheres. A similar survey has yet to be attempted for the Earth.

Streaks are prominent features of aerial and space images of the Central Namib (figures 2, 40-43), and occur also to east and south of the dune sea. Occurrences of wind streaks in the Central Namib have been mentioned in passing by Breed et al. (1979) and Fryberger (1979). Streaks of the bedform type appear to be absent in the Central Namib, where they comprise instead (1) thin veneers of particles swept off white marble ridges and dark granite domes, and (2) patterns of deflation of preferred grain sizes. Fine grains produce brighter streaks, and coarser grains darker streaks (Greeley and Iversen 1985). Shear stress is locally many times greater (even an order of magnitude higher) beneath horizontal, obstacle-generated vortices than that exerted by the ambient wind (Greeley and Iversen 1985).

Analysis of streak orientation in the Tumas basin shows that wind streaks are without doubt generated by the strongest flow of the Berg Wind, in this area regularly a northeast wind. Tables in Lancaster et al. (1984) show a preponderance of strongest winds from the northeast and ENE sectors (034°-079°) for all six plains stations (Zebra Pan, Ganab, Vogelfederberg, Gobabeb, Swartbank, and Rooibank; figure 1), and although winds blow from other sectors, especially the southwest, these winds appear to be too weak to leave an imprint on the plains landscapes. Even resultant sand-flow directions that incorporate all wind components are heavily dominated by the northeast component and give resultant sand-moving winds from 065° and 069° (figure 39).

These orientations accord closely with a mean wind streak alignment of 055°, assessed from satellite imagery (n = 28). Values range between 047° and 063°. Mean alignment of a largely different set of streaks taken from aerial photographs was similar (058°, n = 19), all values falling likewise within a narrow range (053°-066°). Streaks with other orientations are not observed except in a narrow coastal tract. Within the east-west Tsondab and Tsauchab river valleys, topographic control exerts a progressive east-west

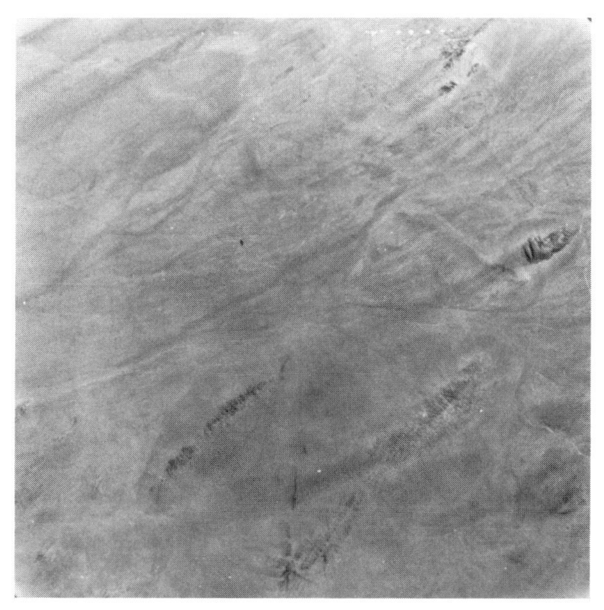

Fig. 40. Wind streaks generated in the lee of various topographic obstructions, e.g., the granite dome (center right) at B (fig. 4) on the Tumas River alluvial plain. Streaks clearly trail to southwest of obstructions. North at top. The dome is 750 m in length. (South African Trigonometrical Survey Aerial Photograph no. 294/4409, with permission)

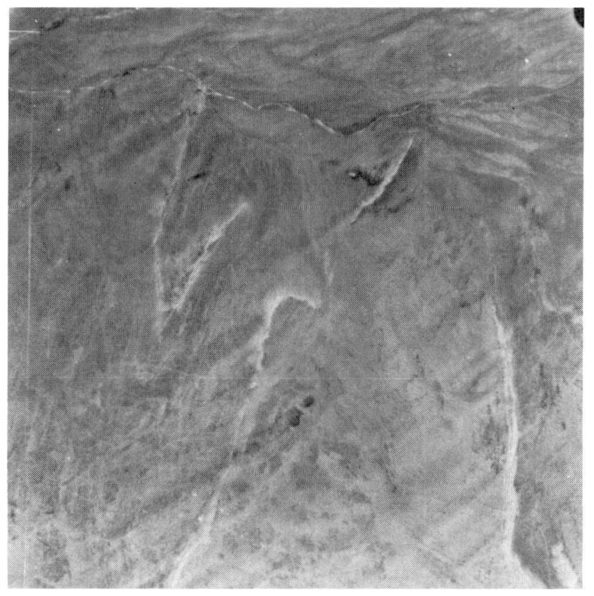

Fig. 41. Wind streaks developed to southwest of marble ridges in Tumas valley sector B-C (fig. 4). North at top. Six kilometers separate the marble ridges abutting the Tumas River. (South African Trigonometrical Survey Aerial Photograph no. 507/3899, with permission)

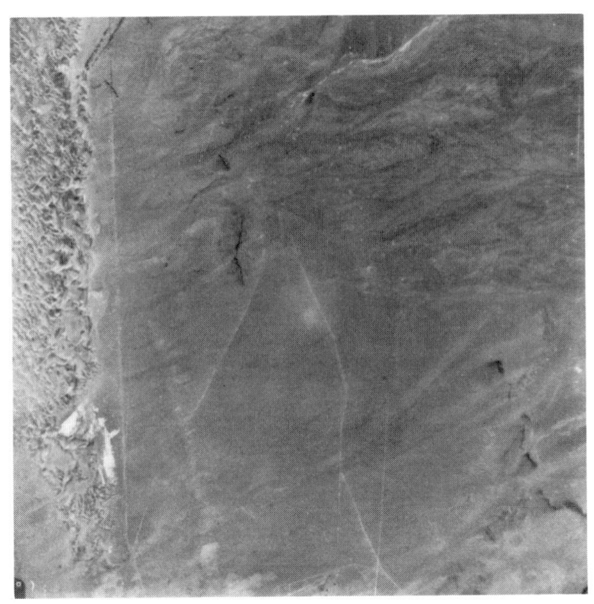

Fig. 42. Wind streaks developed southwest of dolerite dikes (black angular lines) near Tumas Vlei (lower left). Walvis Bay-Swakopmund coastal dune cordon along left margin. North at top. Vlei measures 2.5 km in N-S directiion. (South African Trigonometrical Survey Aerial Photograph no. 760/5005, with permission)

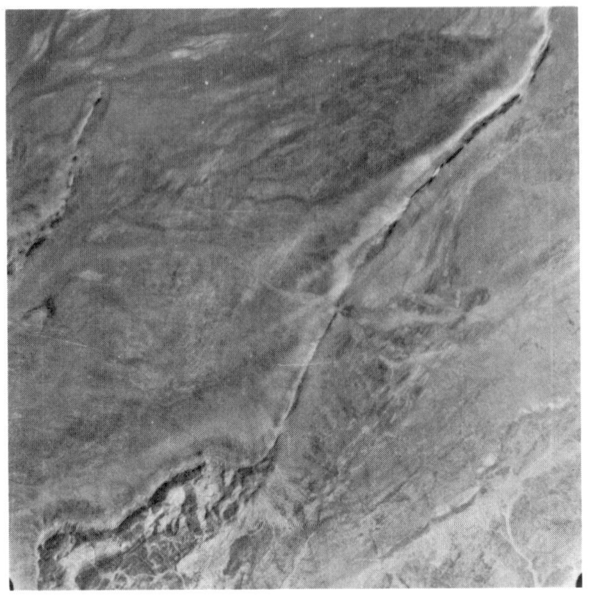

Fig. 43. Wind streaks trailing southwest of Marmor Pforte Ridge on Welwitschia Flats. North at top. Scale: photo 14 km on a side. (South African Trigonometrical Survey Aerial Photograph no. 507/4579, with permission)

trend on wind streaks toward valley centers. Ground observation leaves no doubt of coppice and ephemeral dune development during Berg Wind events. Dust plumes are generated by the Berg Winds (Jaeger 1965) and move offshore in a southwesterly direction (e.g., METEOSAT visible image, June 13, 1979, 13h30).

Berg Wind Features in the Dune Sea

Lancaster (1985) has commented on the unusual situation that sand-moving winds on the plains are effectively dominated by northeast winds whereas the dune sea is dominated by SSW winds—a narrow (5 km wide) transition zone at the north end of the dune sea apparently separating the two. Lancaster (1985) suggested that the "mesoscale roughness of the dunes acts to reduce the velocity of northeasterly winds." But the situation seems "anomalous" (Lancaster 1985:618) indeed, since the winds involved are not simply local, topographically induced winds. They are part of deeper land-sea and geostrophic systems (Tyson and Seely 1980; Lindesay and Tyson 1990).

It is suggested that sand-flow resultants (e.g., figure 39) may not always be useful in appreciating the work of wind. It is reasonable to suppose, for example, that these opposing winds produce eolian effects on both sides of the Kuiseb River. Ward and von Brunn (1985) have documented winter immobility of north-moving dunes in the Kuiseb delta as a result of the effect of the Berg Wind; and lack of sand and dominance of the Berg Wind negate any geomorphic effects of the SSW wind on the flats north of the Kuiseb River.

South of the Kuiseb River the opposite is not the case, however. Here a plentiful supply of movable sand exists well inland of the coastal tract. Indeed, Berg Winds appear to generate a variety of effects across the entire width of the dune sea. These effects are probably both permanent and seasonal, except for the narrow coastal strip of transverse dunes where exceptionally strong southerly winds obliterate any Berg Wind features—even though Berg Winds are felt on the coast, as records from coastal locations show (Logan 1960; Jackson and Tyson 1971; Ward and von Brunn 1985).

The features that appear to be most securely related to the Berg Wind are the small linear dunes, termed "corridor crossing dunes" by Lancaster (figures 38, 44). These features have been mentioned in descriptions of the dune sea. Crossing dunes are generally small (6-8 m high, Lancaster 1980) (figure 44) and usually do not extend beyond one interdune corridor. That is, they are markedly shorter features than the large linear dunes and usually significantly lower. Where best developed in the central tract between Tsondab and Sossus vleis, however, the crossing dunes bulk almost as large as the north-south dunes. Besler (1984) has termed the resulting pattern a

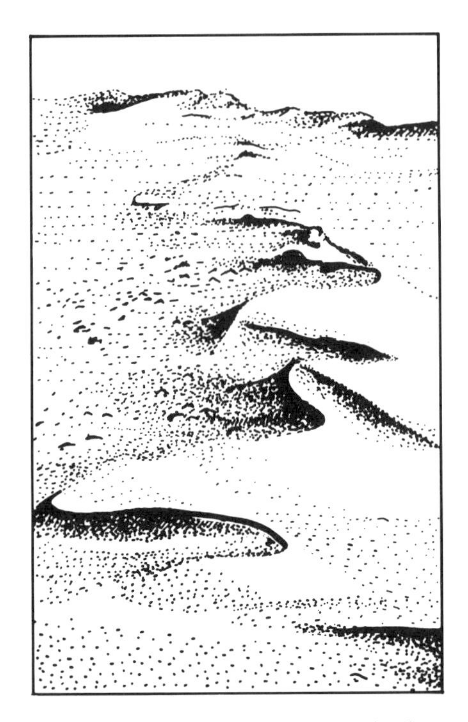

Fig. 44. Corridor crossing dune (1-2 m high in northern Dune Namib (from a photograph by E. D. McKee 1982).

"network complex." Here, individual crossing dunes span three or four corridors before dying out.

The northeast-southwest orientation of the crossing dunes is notably consistent. Yet the precise correspondence of crossing dune alignments with those of the wind streaks, and hence of maximum Berg Wind flow, has not been remarked upon, probably because of the compelling correspondence between very strong, well-documented southerly winds at the coast, and the striking north-south alignment of the dominant dunes in the dune sea.

That the streak alignments coincide with those of the crossing dunes is evident (figure 38). The mean azimuth of a population of 149 crossing dunes in the northern half of the dune sea is 057.56°, ranging between 045° and 074°. Compared with wind streak means of 055° and 058°, there can be little doubt that the Berg Wind is the formative wind for these features (compare figures 40 to 43, and 45 to 47). No statistical difference was found between streak and dune populations. Lancaster (1983) mentioned that secondary arms of star dunes in the eastern tract of the dune sea are also elongated northeast-southwest.

Another reason for the difficulty of ascertaining the elongation direction of crossing dunes is that points of origin are usually impossible to determine. Some dunes peter out at their northeast extremities, some at their southwestern, and many connect at both ends with the large north-south dunes. By contrast, wind streaks usually display clear points of origin at topographic irregularities such as ridges and granite domes (figures 40 to 43). Southwesterly extension under the influence of Berg Winds is thus not in question for the streaks, nor therefore, by inference, for the crossing dunes. One small crossing dune gives direct support for the hypothesis presented here. Located immediately south of Sossus Vlei, this dune is attached at its northeast end to the pointed summit of a larger dune; it disappears 500 m to the southwest. There is little doubt that the crossing dune has extended from the northeast under the influence of the Berg Wind.

Discussion

It is concluded that Berg Wind flow regularly crosses the entire width of the Dune Namib, generating various eolian features, the most notable of which are the crossing dunes. These are recognized by their precise directional correspondence with wind streaks north of the dune sea. Crossing dunes are encountered throughout the northern dune sea and at places in the southern half, but not along the narrow coastal strip in which high energy, southerly winds apparently remove such features.

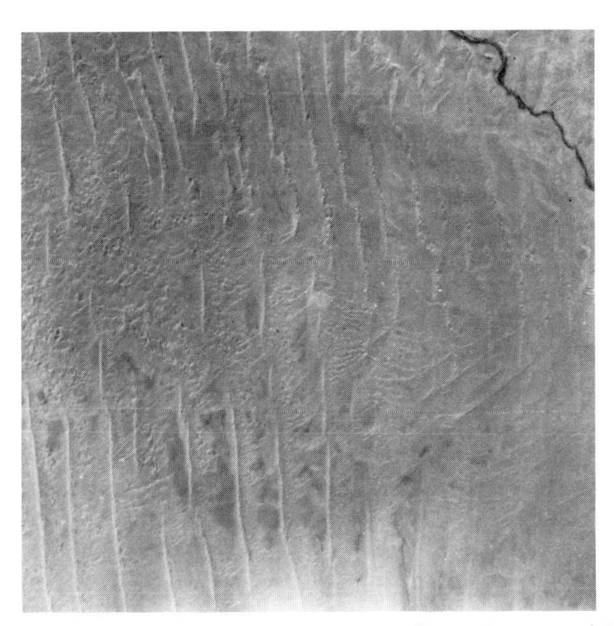

Fig. 45. Northern Dune Namib dune patterns: major linear dunes trend NS, with well-developed crossing dunes, trending NE, center right. Kuiseb-Sout river confluence top right. North at top. Major dunes, bottom center, 2.5 km apart. (South African Trigonometrical Survey Aerial Photo no. 776/1064, with permission)

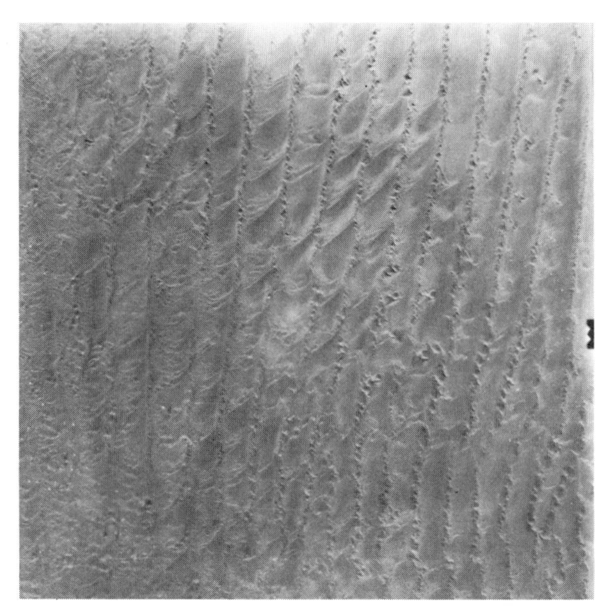

Fig. 46. Reticular pattern of major and minor linear dunes near Sossus Vlei, illustrating an area of well-developed corridor crossing dunes (NE-SW aligned). Detail of outlined area shown in figure 47. North at top. Scale: major dunes NE quadrant 2.4 km apart. (South African Trigonometrical Survey Aerial Photo no. 776/1066, with permission)

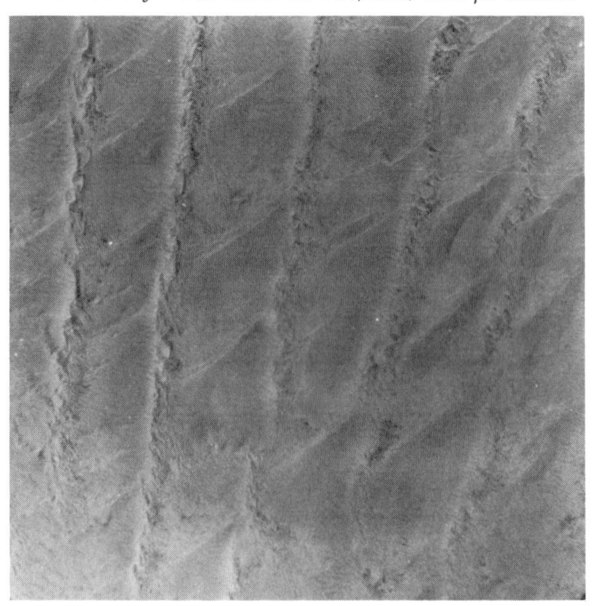

Fig. 47. Detail of figure 46. Points of origin of crossing dunes generally unclear. North at top. Scale: widest dune street 2.6 km wide. (South African Trigonometrical Survey Aerial Photo no. 663/5320, with permission)

Referring to his work on oblique propagation of dunes, Tsoar (1978: 141) concluded: "No proof was found that the longitudinal dune is formed by the helicoidal flow or by a unidirectional wind." He stated that "oblique cross winds are the main factor required for longitudinal dune dynamics and morphology" (Tsoar 1978:141). The data presented above suggest the opposite, that unidirectional and nonoblique winds do indeed generate linear dunes, at least of crossing-dune type in the Namib sand sea.

It seems unnecessary to invoke complex effects such as oblique propagation of linear dunes when good evidence exists for a direct relationship between crossing dunes and wind streaks. Indeed, many researchers have favored the model of linear extension parallel to the resultant wind (Bagnold 1953; Cooper 1958; Wopfner and Twidale 1967; McKee and Tibbetts 1964). Greeley and Iversen (1985) have suggested that the oblique propagation model may not be applicable in all deserts.

The fact that dominant sand-moving winds in the northern Dune Namib are SSW and northeasterly calls into doubt the geometries of Lancaster's theory of evolution of the major dunes: it seems unlikely that a bimodal regime, comprising these vector maximums, could generate linear dunes extending due north, or even NNW, against the stronger, northeast winds (see below), as claimed by Lancaster (1982a, 1983).

Lancaster (1980) reported the existence of a field of small barchans on the Tsondab Flats with slip faces oriented northeast. Some barchans display strongly elongated southern horns, which Lancaster (1980) ascribed to the effect of bimodal winds from the SSW quadrant. The interpretation offered here, however, is that the elongated arms are crossing dunes, since they are aligned precisely with other crossing dunes. It appears therefore, that SSW winds have subsequently caused recurving of the southwest end of crossing dunes, imparting a barchanoid form. Although this explanation is not entirely satisfactory, the alternative scenario (Lancaster 1980) does not counter the great weight of evidence in favor of the present efficacy of the Berg Winds.

It is possible to envisage a synchronic situation whereby relatively strong Berg Winds fashion the crossing dunes, but SSW winds of relatively long duration move sand along them in the opposite direction most of the year. The dune patterns may be thus a product of complex interactions between both sets of winds. East-flank dunes (figure 38) are transverse to both winds and may be predicted to reverse during winter, a prediction borne out by field observation (M. Seely, personal communication).

Models related to such multiple sets of formative winds have been proposed for some dune fields (Cooke and Warren 1973), but are as yet generally too complex to model effectively, since the dynamics of dune formation remain so poorly understood.

A diachronic theory indeed seems simpler. Berg Winds are actively realigning fossil dune forms into conformity with arguably the strongest modern flows. Though not mentioning crossing dunes, Besler (1975, 1980, 1984) has suggested that the east-flank barchanoids, linear northwest-oriented silk dunes, and small foredunes are manifestations of such remodeling. The small size of these types in comparison with the major linear dunes suggests that they may be relatively young.

Indeed Wilson (1972) has taken it as axiomatic and builds theories of dune development on the proposition that bedform age and size are related: thus, ripple bedforms are younger than larger forms such as dunes, and dunes are younger than draa-sized features. In the same way, it seems plausible that crossing dunes are younger forms than the major, north-south dunes, since they are on average an order of magnitude smaller than the latter. The existence of the crossing dunes is thus suggestive of a younger dune set formed by the resculpture of older dunes.

By Besler's theory, stronger winds are required to account for the spacing of the large linear dunes. Evidence appears to be accumulating that wind velocities indeed increased during the Last Glacial Maximum. Nicholson and Flohn (1980), Newell et al. (1981), and Flohn (1984), among others, have argued that Glacial Maximum wind speeds were higher than present on the grounds of steeper hemispheric temperature gradients.

Sand-flow data from the Plains Namib support these conclusions. Calculations show that annual potential sand flow for the noncoastal areas of the northern dune sea and neighboring plains (Lancaster 1985) is more than twice as high today under Berg Wind conditions as it is under SSW wind conditions: at the plains station RB (Rooibank) a total of 278 tonnes/m/yr. (resultant 129 tonnes/m/yr.) is moved, compared with 23-119 tonnes/m/yr. (resultants 6-63 tonnes/m/yr.) at stations in the northern and central dune field.

Furthermore, dust plumes are generated by the strongest Berg Winds once or twice per year, and cross the west coast in a southwesterly direction (e.g., METEOSAT visible space image, June 13, 1979, 13h30). Examination of two years of METEOSAT daily images showed no dust plumes generated by southerly quadrant winds despite the existence of vigorous, daily winds from the south. Since the Berg Wind is apparently constructing crossing dunes at present, it may be true that winds with the strength of present Berg Wind maximums were required for the formation of the north-south dunes.

Inferences from dune spacing further support the contention that stronger winds were operative in building the north-south dunes. Wilson's (1972) opinion that stronger wind speeds may relate to greater linear dune spacing lends prima facie support to Besler's arguments. Furthermore, it is

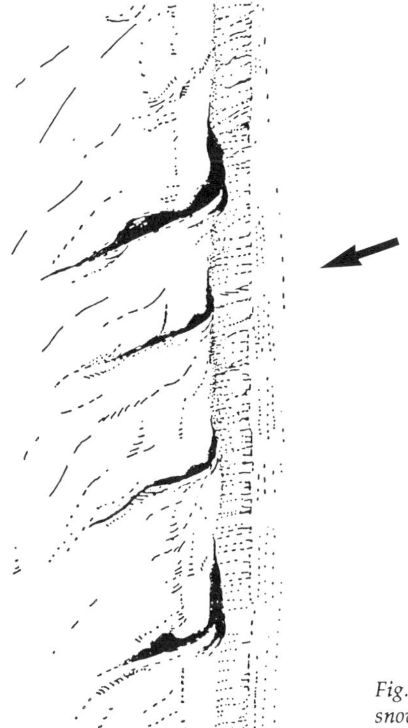

Fig. 48. Regularly spaced linear snowdrifts downwind of obstruction (ditch) (after Iversen 1979).

noteworthy that the spacing between crossing dunes, where sets of crossing dunes have developed, is the same as that of the major dunes. It has been shown that the Berg Winds are stronger than, and presently move more sand than, the SSW winds, suggesting that winds with higher velocities—of the order of present Berg Wind maximums—may have been necessary to establish the spacing of the north-south dunes.

The diachronic view helps explain another aspect of dune patterning, namely the regularity of spacing of the crossing dunes. Smaller aerodynamic forms (such as crossing dunes) are known to achieve regular spacing when air flow crosses an elongated obstacle. Under such circumstances transverse but cellularly-organized helices develop along the lee side of the obstacle (Maull and East 1963). Since the cells tend to develop to a constant size, erosional and depositional forms associated with the cells are regularly spaced. Iversen (1979) has noted the application of this phenomenon of fluid dynamics to the regular spacing of linear snowdrifts (figure 48).

The significance of these observations is that they appear to explain both the magnitude and regular spacing of the crossing dunes, the Berg Wind viewed as flowing approximately transverse to the elongated obstacle of the north-south dunes. More important to the present argument, and even allowing that the SSW winds may be formative of the crossing dunes, the spacing of the crossing dunes implies the prior existence of the north-south dunes: without the requisite obstacle in the form of the north-south dunes, the regularity of crossing dune spacing is less easily explained.

Lancaster (1982a) has objected to the helicoidal theory on the grounds that helices of the small size required to account for Namib dune spacings do not exist. Since evidence of horizontal vortices of a variety of sizes has been forthcoming (Angell et al. 1968), the objection falls away.

Conclusion

One of the more secure facts available in discussions of linear dunes and formative winds in the central tract of the Dune Namib is the apparently causal relationship—argued for in this study—that exists between northeasterly Berg Winds and crossing dunes. From implications of this connection, and various theoretical and paleoclimatic considerations, it seems likely that except for a thin, mobile skin of sand, the large linear dunes may well be fossil features, as Besler has argued.

For the present study, these conclusions mean that southerly winds, stronger than present, affected the Tumas basin and northern sand sea during the late Pleistocene, and probably therefore during prior, glacial episodes. This model accounts for the replacement of present winds by stronger sand-moving winds in the interior of the Namib.

Stronger winds also explain the spacing of the dunes, and their postulated age accords better with dune size. Furthermore, expansion or shifting of the South Atlantic gyre in an onshore direction would replace the modern oblique SSW winds—arguably too weak to give rise to the larger linear dunes—first by stronger winds, and second by winds aligned more closely with the extant dune pattern. Such winds conceivably might have provided a resultant more effectively parallel to the dune pattern. The diachronic theory at present accounts for more of the observed phenomena than does the synchronic.

More energetic winds that have been postulated would undoubtedly have affected the plains to the north of the dune sea, although today all features that such winds may have generated appear to have been obliterated. Two exceptions may exist: first, a small percentage of boulders on the plains near Gobabeb that display faceting oriented to southerly winds (Selby 1977b); second, the coastal dune cordon between the Kuiseb and Swakop river mouths.

In an attempt to gauge how long the Berg Winds may have acted as formative winds in the Central Namib, it seems relevant that ventifaction, which is strongly dominated on the plains by these winds, probably takes thousands of years to accomplish (Higgins 1956). Besler (1976) and Selby

(1977b) have suggested that the Berg Wind has been faceting boulders for much of the Holocene. Psammophilous rodents, common in the dune sea, are not found in 8000-year-old owl pellet deposits from shelters north of the Kuiseb River. This fact led Brain and Brain (1977) to conclude that dunes have not existed in this area of the plains during the Holocene at least.

Chapter 9

CHRONOLOGY AND SYNTHESIS

Dead water and dead sand
Contending for the upper hand.

—T. S. Eliot, *Four Quartets*

Seaduma se duma ka mor'a sekgutlo
[Thunder rumbles at the back of the valley]

—D. P. Kunene, Heroic Poetry of the Basotho

Problems of interpretation and external correlation are ever-present in the kinds of reconstruction undertaken in such a study as this, but the Tumas data holds potential for progress in several spheres. This chapter presents a paleoenvironmental and chronological synthesis of evidence from the Tumas basin. I have linked the study-area chronology with those of other areas where warranted, recognizing the dangers of such correlations when absolute dates are not available.

Discussion centers on three broad geomorphic regimes, in each of which the nature of the evidence and the trends differ. The earliest period concerns aggradational environments responsible for deposition of the Leeukop and Tumas Formations. Within this period trends of probable regional extent can be discerned, from hyperarid eolian to arid or semiarid fluvial morphogenetic environments. The second concerns a fundamentally different period in which stages of soil formation are punctuated by stages of river incision. The youngest period incorporates features of probable Holocene age and a host of small, subrecent to modern features.

At minimum, the following chronology can be identified in the study area (table 5). Individual events differ in their usefulness for paleoenvironmental reconstruction. Sequences of events per se are sufficiently detailed to encourage the establishment of correlations external to the Tumas drainage (table 1). Dating of events is frustratingly inadequate throughout the Tumas and neighboring drainages. Plausible dates, consistent with the

TABLE 5.

Chronology of events and inferred environments in the lower Tumas River basin

	Event	Paleoclimate
13.	Mobilization of gypsum and uranium surficially on all incoherent materials; stream bank colluvia; Berg Wind–related wind streaks; minor incision (surface S4). Holocene. (Southerly winds dominant pre-Holocene?)	hyperarid (rainfall 0-50 mm/yr.)
9-12.	2d and 3d phases of gypsification, incision, and micropedimentation (surfaces S2, S3).	hyperarid with moist-episodes (up to 200 mm/yr.)
8.	1st incision by Tumas stream. Lower to Middle Pleistocene.	moister (up to 200 mm/yr.)
7.	Tumas sediment depository recedes; gypsification of surface S1.	arid (rainfall < 200 mm/yr.)
6.	Multiphasic emplacement of uranium mineral by high water tables; Tumas Sandstone Mb. 2 gravel sheet (depositional surface S1). Miocene at youngest.	arid-semiarid fluvial activity: increased rainfall on escarpment
5.	Onset of permanent upwelling (post mid-Miocene). (Gypsification?)	arid
4.	Tumas Sandstone Mb. 1 (red sandstone). Miocene.	arid-semiarid fluvial activity: increased rainfall, on escarpment at least.
	(Dune sand accumulation. Eocene-Miocene?)	(arid)
3.	Lower Swakop/Khan drainage reorientation. Incision by proto-Swakop; cutting of Tumas buried canyon shoulder.	endoreic drainage
2.	Goanikontes fm. (Eocene); Leeukop Conglomerate Fm.	arid to semiarid
1.	Namib Unconformity Surface (NUS) cut as rugged valleys (Tumas canyon). Lower Paleocene or Oligocene.	probable exoreic drainage

regional geological succession and estimations of Martin (1950, 1961), Korn and Martin (1957), and Ward et al. (1983), are tentatively assigned.

1. Valleys, including the buried Tumas canyon, are cut into the Namib Platform by the exoreic proto-Swakop and proto-Khan rivers. Lower Paleocene or Oligocene.
2. The relatively thick (90 m) Leeukop Conglomerate Formation is deposited as a valley fill. Paleocene or Eocene.
3. Erosion widens the upper levels of the buried Tumas canyon, producing a marked bench near the upper boundary of the Leeukop Formation. Reorientation of the lower Swakop River occurs latterly.

4. A red sandstone, Member 1 of the Tumas Sandstone Formation, 10 to >35 m thick, is deposited in a long series of mass flow-related deposits on the distal alluvial plain. Miocene. Prior dune sand accumulation, in the central Tumas basin at least, is strongly implied. Eocene to Miocene.
5. Gypsification of the surface of the Tumas Formation takes place and authigenic iron oxide release alters the color of the sandstone probably from this stage onward. Post–mid-Miocene oceanic upwelling is established.
6. Thin sheet gravels and sands are laid down across the alluvial plain as Member 2 of the Tumas Formation, giving maximum expression to a long-profile convexity in the lower Tumas valley (upper surface of this alluvial body termed surface S1). Miocene-Pliocene.

   Valley widening, related approximately to the final level of alluviation, proceeds unhindered from this stage. High water tables characterize this and several subsequent, undated periods, as evidenced by a thick hydromorphic zone in the red sandstone, and by emplacement of the soluble uranium mineral carnotite, also on more than one occasion.
7. Gypsification of surface S1 commences. Recession of the Tumas sediment depository occurs. Miocene to Upper Pliocene.
8. The Tumas River incises into S1 by approximately 10 m and flanking micropediments (S2 surfaces, up to 0.5 km wide) duly evolve. Lower to Middle Pleistocene.
9. Gypsification commences on S2 surfaces and continues on S1.
10. The Tumas River incises for a second time by 1-3 m with extension of related S3 micropediment surfaces. Middle to Upper Pleistocene.
11. Gypsification commences on S3 and continues on surfaces S1, S2.
12. The Tumas River incises a third time (ca. 1 m) giving rise to associated micropediments, the S4 surfaces. Upper Pleistocene.
13. Gypsification commences on S4 surfaces, and continues on S1-S3. Minor incision by the Tumas River (<1 m) occurs. Thin sand veneers spread on east-facing hillsides. Coppice dunes and wind streaks form actively. A small dune field develops in the eastern study area. Ventifaction and karren development occur and deflation maintains steep scarplets. Banks of the Tumas River display submodern colluvial mantles with gypsum and carnotite impregnation. Thin gravel spreads, as terminal gravity flows, are laid down. Holocene, submodern.

The Namib Unconformity Surface

In chapter 3, I attempted to fix the Tumas depositional events in the regional succession. Likely ages were based on subcontinental manifestations of eustatic sea level movements synthesized for the entire coastline of southern Africa by Dingle et al. (1983). Two of the more firmly fixed events in an unfirm set are (1) deposition of the Goanikontes lagoonal sediments that seem best related to Eocene transgressions, and (2) enhanced discharge in the Tumas River coincident with pan-Namib fluvial episodes in the Miocene. Subsequent incision phases seem to parallel similar, poorly dated episodes in the Tsondab watershed.

Valley cutting, the first stage of the Tumas chronology, is either pre-Eocene or early Miocene in age. It has been argued that the former is more probable. Valley cutting suggests that runoff from the plateau via the proto-Swakop and proto-Khan rivers was sufficient then, as in the recent geological past, to enable these rivers to perform significant incision by maintaining connection to the sea.

The valleyed topography of stage 1 represents the discontinuity between ancient rocks and the Tertiary Namib Group rocks, a contact termed the Namib Unconformity Surface (NUS). The NUS is significantly flat in other areas, but it has been shown that the buried feature in the Central Namib is by no means flat. Sculpture of a rugged coastal foreland, in an area where major rivers debouch, seems plausible; flat surfaces planed across the ancient rocks in the Tumas basin are apparently related to topographic levels established by the final phase of alluviation (stage 6). Such relatively recent sculpture (or rebeveling of prior planed landscapes) is responsible for the present, exposed NUS in the lower Tumas basin, making it a youthful rather than Late Cretaceous feature, as it appears to be in the southern Namib Desert.

Since ancient, buried relief is rugged, and young relief is planed in the Central Namib, the view that the NUS is everywhere ancient, or that it is everywhere a planed feature, must be treated with caution.

Because of the controversy concerning climatic control of the development of features such as pediments, long slopes of low declivity, and granite domes (see for example, discussion in Busche and Hagedorn 1980), it is considered that the existence of such features in the Tumas basin represents an inadequate basis for the evaluation of Namib paleoenvironments. Büdel (1977:203-204) summarizes the thinking of national schools concerning formative environments of these features thus: "Many researchers, especially in France, Britain, and America, [assume an] early advent of full desert conditions . . . as far back in the Tertiary as possible. . . . Some advocates of this view go so far as to regard the entire Saharan etchplain and inselberg relief as a product not of a former humid tropical climate, but of a long-term and still active desert climate."

This controversy is evident in comparing the works of Rust (1970) and Selby (1977a) on Central Namib landscapes. Rust was unable to contemplate evolution of the bornhardt landscape in the Komuanab area, 100 km northeast of the study area, without invoking Tertiary deep weathering. On the other hand, Selby (1977a) concluded that granite domes in the upper Tumas valley and neighboring parts of the Inner Namib are structural inselbergs surrounded by schist pediments, a landscape that in his opinion does not presuppose an origin by deep weathering. There is no direct evidence of deep weathering near the surface on the Tumas or Welwitschia Flats. Equally, rock-cut surfaces in these areas may represent stripped basal surfaces (the lower of the "double planation surfaces"; Büdel 1957:38). This study favors neither argument. Arguments concerning pedimentation do not apply to the cutting, in weak lithologies, of features referred to herein as micropediments, the *glacis d'érosion* of Dresch (1957).

Depositional Stages: Early to Mid-Tertiary

Leeukop Conglomerate

The Leeukop Formation, stage 2 in the Tumas chronology, comprises coarse, probably bed-load material with few fines. These textural considerations, as well as the manifestly discontinuous bedding of the Member, are suggestive of semiarid fluvial environments in the lower valley of the proto-Swakop River. The Leeukop Formation is probably a continental correlative of early or mid-Tertiary global transgressions.

The hiatus at the top of the Leeukop Conglomerate Formation gravels, inferred from the buried shoulder in the Tumas canyon walls (stage 3), indicates reduced local deposition. This event has no necessary paleoenvironmental significance because exoreic rivers—such as the Swakop may be assumed to have been—do not necessarily alluviate their beds, being sediment-passing systems.

Implied Phase of Dune Activity

Deposition of the red sandstone as Member 1 of the Tumas Formation (stage 4) may have taken place at any time after deposition of the Leeukop Conglomerate and reorientation of the proto-Swakop River.

It has been shown that the Tumas Sandstone Formation comprises a long series of mainly sediment gravity flows, features common in deserts, mainly in alluvial fan settings. Gravity flows comprised of fine sands lacking gravel-sized material are less common, since viscous flows are characteristically competent to transport very coarse clasts. It has been argued that the marked predominance of one particular textural class implies prior (extrinsic) control of sediment caliber, since scattered coarse inclusions give manifest evidence of high competence. Because the modal class of particles in the sandstone lies squarely in the narrow range of world dune-sand textures, it has been suggested that blown sand was the source material for the great number of flows that apparently make up Member 1 sands.

The clear implication is that a dune field existed in the central Tumas basin. That remnants of the original dunal structure appear not to be extant suggests substantial reworking of the eolian sediments by fluvial processes, and deposition with fluvial admixtures, in the form of the red sandstone. The existence today of a small dune field and widespread pebbly and cobbly mass flows on the lower Tumas alluvial plain adds circumstantial evidence to the argument. The volume of the Member 1 sandstone suggests, however, that a substantially larger field of sand was reworked.

The common association of dunes with nearby deflatable river sands (Reineck and Singh 1980), and especially with distal arid basins, suggests that the dune sands themselves were most likely locally derived from fluvial deposits on the Tumas Flats. The implied volume of blown sand suggests accumulation over a significant period of time. Although the evidence is tenuous, the present small dune field has probably been in the process of formation for the duration of the Holocene at least. A substantially longer period of time is implied to account for the accumulation of sand in the postulated dune field as a distinct epoch between stages 3 and 4.

Furthermore, Besler and Marker (1979) and Ward et al. (1983) have shown that the Tsondab Sandstone was more widespread than it is now. On the strength of the pan-Namib stratigraphic sequences as presently known, the Tumas dune field may well have been an outlier of the Tsondab Sandstone, probably derived by deflation of local alluvia. Besler (1984) has postulated a phase of fluvial reworking of dunes in the large dune field south of the Kuiseb River, suggesting that this kind of interaction is not unique to the timespan of Member 1.

Since significant eolian activity currently takes place to west and south of the study area, climates responsible for the dune field were probably as arid as today's. The volume of the sands implies persistent aridity.

Period of Fluvial Activity

An extended period of fluvial activity, sufficient to accomplish reworking of sand in the dune field into a sedimentary body 4 km wide, 37 km long at minimum, averaging 20 m in thickness, by deposition of numerous, thin, highly discontinuous flows, is suggested by the existence of the Member 1 red sandstone. With no evidence of soil development between flows, and occasional gravel influxes as a result of particularly strong flows, a period of increased and sustained fluvial activity—certainly increased above present levels—is also suggested. A sufficient number of these discharge events passed the valley constriction at point B (figure 4) to give rise to the almost 30-km-long convexity in stream profile (between points A and C, figure 4), to which so many signs point.

The existence on the alluvial plain of gravity flows, the sheetlike form of the sandstone at both architectural and alluvial-plain scales, and the indication of a localized convexity of alluvial surface all place the paleosetting squarely with arid endoreic environments.

Two observations seem important. First, neither episode of stage 4 can be conceived of as a single, possibly catastrophic event. The impression is quite the opposite, that is, of a significant length of time required to accomplish the work first of eolian, and then fluvial deposition.

The second observation concerns the apparent sensitivity of the record to what were patently small, though apparently persistent, climatic shifts. These shifts nevertheless imposed dramatic morphogenetic changes in the lower Tumas basin, dune fields giving way to alluvial plains. Both environments are notably arid in character.

It must be postulated that the escarpment zone, at least, became wetter than it is today, probably semiarid, in order to explain the existence of mass flows, considering that stream discharge is today almost nonexistent in the Tumas River. That gravity flows in fluvial settings are features generated by flow cessation indicates that such discharges regularly ended on the lower Tumas plain. The increase in stream flow in the Tumas River must have been just sufficient to promote the evacuation of sediment, indicating climates no more than semiarid along the escarpment, and though slightly moister, remaining arid in the western Tumas basin.

Preservation of the Tumas Formation, which is arguably Miocene in age, supports the contention that exoreic discharge has been nonexistent, or so insignificant in the Tumas catchment that Member 1 beds have remained extant in the alluvial plain since deposition. Stream erosion usually removes valley-floor debris flows (Bull 1977) in climates associated with stronger or more permanent stream flow. It has been noted that local convexities are features of the long profiles of streams in the vicinity of floodplain constrictions in the Namib Desert. The existence of the large convexity in the lower Tumas is further evidence for aridity during construction of the feature, in a river tract in which discharge diminishes.

Parallels with the Tumas Sandstone Sequence

Stage 4. Support for the inferred existence of a prior dune field and for interpretations of events in stage 4 comes from neighboring drainages to the southeast and south, where strikingly similar landforms and chronologies exist, suggesting that patterns of past formative environments are becoming evident. In both the Ubib and Tsondab drainages, 140 km southeast and 100 km south of the study area respectively, the sequence begins with dune sand accumulation, in the form of a 100-200-m-thick eolian body in the Ubib valley (Hüser 1976), and in the form of the widespread, 60-200-m-thick, tabular mass of the Tsondab Sandstone Formation, also dominantly eolian (Barnard 1973; Besler 1977; Besler and Marker 1979; Ward 1984). The Tsondab Sandstone is a major Tertiary unit that underlies the entire Namib sand sea in the Tsondab drainage area, continuing unbroken into four drainages further south.

The arid episode signified by the existence of eolian sands may have been more arid or of longer duration than the recent episode under which the present dune field has evolved. If the sand sheets represented by the Ubib, Tsondab, and Tumas occurrences are coeval, they represent eolian environments that stretched further inland than the present dune field, since they underlie presently calcreted and vegetated parts of the most humid Inner Namib. Regardless of the putative existence of the Tumas sand field, the other sheets undoubtedly imply more intense aridity and/or a longer period of aridity than is responsible for the presently active dune field.

The second stage in the Ubib and Tsondab valleys is fluvial reworking of the dune-sand surface. In the case of the Tsondab Sandstone, there is no doubt that the upper surface of the Formation has been entirely modified by flowing water. Many researchers have commented in passing on what must be regarded as a crucial phase of morphogenesis. Ollier (1977:208) described the surface as a "vast pediment," which he termed the Tsondab Planation Surface. Besler (1980, 1984) has documented the constancy of declivity of the surface from the escarpment almost to the coast. She concluded that the surface had been fashioned by fluvial processes with a corresponding depositional wedge nearer the coast. Furthermore, the Planation Surface is associated with extensive alluvial gravel spreads (Besler 1980, 1984; Lancaster 1984b). This pattern occurs not only in the Tsondab drainage, but also in the four drainages to the south.

Likewise, Hüser (1976) argued for planing of a sand sheet surface in the Ubib catchment by flowing water (figure 49), with subsequent slope activity of such vigor that notably coarse detritus was transported across the new surfaces.

The similarity of evolution, involving at minimum the complete refashioning of the surface of a sand sheet, is echoed in a third basin, namely the lower Tumas. The parallels of significant blown sand accumulation, probably from immediate sources, and the morphogenetic shift to significant fluvial remodeling, suggest a sequence of events of possibly regional importance.

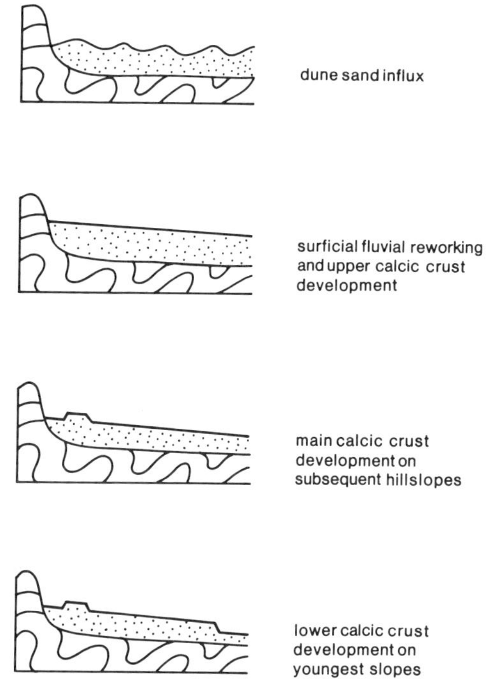

Fig. 49. Sequence of geomorphic events in Ubib River drainage (after Hüser 1976).

As with the Tsondab Sandstone, dated anywhere from Eocene to Miocene (Ward et al. 1983), eolian sand accumulation in the lower Tumas basin could be of equivalent age. It has been argued that the Tumas Sandstone is coeval with the fluvial period that followed eolian accumulation in both the Ubib and Tsondab watersheds. Present opinion favors a Miocene age for this fluvial period (Martin 1950, 1961; Ward et al. 1983).

Stage 6. The deposition of Member 2 of the Tumas Formation, a relatively thin sheetlike gravel and gravelly sand (stage 6), is interpreted as the product of predominantly unconfined sheet floods. Member 2 consequently suggests semiaridity but may relate to fully arid fluvial activity. Although it is thin, its extent across the entire surface of the lower Tumas alluvial plain suggests a not inconsiderable geomorphic phase of deposition, certainly more than a single catastrophic event.

Furthermore, it finds repeated parallels to the southeast in the other two drainages. In the Ubib sequence, the smoothed surface is capped by a <3-m-thick unit of coarse rubble (Hüser 1976; Blümel 1982) and by the 2-5-m-thick "capping conglomerate" (Lancaster 1984b) in the Tsondab sequence. Thus, even if the event were short-lived or unrepresentative of prevailing climates, neither of which possibilities appears likely, the presence of the unit lends credence to a stratigraphic sequence of regional extent.

It is possible, but less likely considering its great relative size, however, that the Member 2 conglomerate is significantly younger than Member 1—associated with one of the various phases of incision and deposition documented in the Tsondab valley, all of which phases are considered to represent stronger-than-present flow in that river (Lancaster 1984b). The Hamilton Vlei conglomerate on the 80 m bench in the central Tsondab valley, for example, is tentatively correlated with the Oswater Conglomerate in the Kuiseb canyon (Lancaster 1984b).

For the moment, Member 2 is best regarded as the product of the generally moister period of probable Miocene age and pan-Namib extent.

Alluviation in sheetlike form is the mark of Member 2 gravels as well as the capping conglomerate of the Tsondab Sandstone Formation. It has been argued on theoretical grounds that unconfined flows in aggrading fluvial environments, especially with respect to coarse-member deposits (and with respect to deposits that lack differentiation into alluvial plain subenvironments), are probably characteristic of arid sedimentary environments, among others. Member 1 sandstone is likewise deposited across the full width of the alluvial plain. Architecture of this kind, especially that lacking any indication of channelized flow, as in the case of the red sandstone, is strongly suggestive of sheet-flood and gravity-flow alluviation in arid environments.

Water Table Evidence

Evidence concerning water table levels has not been reported from neighboring drainages, but water tables were without doubt high in the lower Tumas basin. Evidence of a thick reduction zone near the surface, when water tables are today 10-20 m, and more, below surface (well below the influence of localized incision in the Tumas constriction convexity), suggests permanently increased groundwater flow, a phenomenon most consistent with increased rainfall in the Tumas catchment. The thickness of both reduction zone and transition zone (between the reduction and overlying reddened zones), as well as the disequilibrium status of the uranium minerals in the Tumas Sandstone, all indicate mobilization or reduction of the minerals involved, under conditions of permanently higher water tables on more than one and probably many occasions.

Such fluctuations occurred after emplacement of Member 1, and probably after emplacement of Member 2. That subsequent incision phases truncate the reduction and uranium-rich zones tends to confirm that the high water tables were associated with earlier stages of the Tumas chronology, though perhaps late (stages 6 and 7) in the early series. High water tables can be plausibly, though tentatively, considered as subsurface manifestations of the Miocene period of increased fluvial activity.

It is reasonable to suppose that thorn bush savanna characterized the lower Tumas alluvial plain during high water table episodes, much as it does in the foothills of the escarpment today. The significant difference is that precipitation falling within the Namib Desert, in the upper, eastern Tumas basin—rather than on the interior plateau—would have elevated water table levels. Tree associations would have become thicker along major arms of the Tumas River itself, similar to those of the present beds of the Kuiseb and Swakop rivers.

Gauging vegetation cover on the surrounding hillsides is more difficult because of the east-west climatic gradient (see below). Cover could have increased above the present very low percentages. The likeliest trend is that the present sparse grasses may have become denser and more permanent, perhaps with occasional, low acacia trees in tributary stream beds during the wettest periods. It is conceivable in the light of present biogeographical patterns that associations of Karoo-like, semidesert scrub may have penetrated from neighboring habitats to east and south during periods of intermediate soil moisture, if these were allied with cooler-than-present temperature conditions. At other times, vegetation probably differed little from the present very low percentages under dwarf desert scrub.

In particularly arid phases such as sand field expansion—whether or not the Tumas basin experienced fully fledged dune invasion—the opposite trend would have occurred, with savanna thornveld retreating from the lower, hotter country up the escarpment. In the lower Tumas basin where aridity cannot effectively be intensified above present levels, very low percentages of vegetation cover will have persisted. One difference may concern expansion of vegetation associations, especially lichens, dependent on fogs that may have penetrated further inland. If fog frequency increased, could vegetation associations have changed with the increased amounts of moisture, in the coastal Namib at least? From palynological evidence in the Transvaal, Scott and Vogel (1983) have suggested that some vegetation associations may have changed in composition under past climatic regimes, rather than simply shifted—unchanged—altitudinally or latitudinally.

The similarity of form and sequence of sediments and landscapes in three separate drainages suggests morpho- and chronostratigraphic equivalence between the drainages. If this is indeed the case, then regional geomorphic, and hence paleoenvironmental, trends are implied: dune sands accumulate in significant thickness surrounding distal desert drainages and probably signify an expansion of the area of hyperaridity northward into the Tumas drainage and certainly eastward into the Inner Namib.

As stream discharge subsequently increases under wetter climates, the dune field is reworked by fluvial agencies, with a final phase of gravel deposition, a marked unit of all three sequences. Greater rainfall locally in the Namib, giving rise to river-dominated depositional environments, is suggested regionally, the fluvial environment arising under conditions that may be described nevertheless as arid or semiarid. Mechanisms are undoubtedly autochthonous and allocyclic (extrinsic), most likely climatic in nature, as Mabbutt (1952) has argued for the Ugab River sequence, Hüser (1976) for the Ubib River sequence, and Lancaster (1984b) for the Tsondab valley landscapes. The morphogenetic shift from great aridity to widespread fluvial sculpture of slopes and valley floors appears to affect the entire Central Namib.

Degradation and Pedogenesis: Miocene to Upper Pleistocene

Morphogenesis in the Study Area

Stage 7 and subsequent stages in the chronology are indicative of a shift in fluvial behavior in the lower Tumas River course, from aggradation on alluvial plain surfaces to incision into them. Morphogenetically, the remaining stages are either erosional in character or quiescent. Where alluviating discharges occurred sufficiently often in the vicinity of the valley constriction at point B (figure 4) in earlier stages—thereby generating the convexity—discharges from the upper Tumas subsequently ceased reaching the constriction altogether. The surface hydrological system appears to have undergone deenergizing to such a degree that deposition was restricted in later stages to the vicinity of point C (figure 4) or yet further upstream.

The system decline, represented by recession of the depository upstream, implies rainfall reduction over the Tumas catchment. Surface sediments of the plain (i.e., Member 2) were rendered inactive and prone to gypsification. Such decline was probably partly responsible for subsequent incision phases, since it implies sediment starvation of flow upstream of the convexity.

Gypsum Crusts. Stage 5 of the Tumas chronology, the equivocal duricrusting event, may mark the beginning of the trend that was to dominate the remainder of the history of the lower Tumas basin. Duricrusting is established fully by stage 7 with gypsification of Member 2 alluvia. Accumulation of gypsum in the supergene environment is one of the surer indications of aridity, since gypcrete crusts are notably soluble in climates wetter than approximately 200 mm of yearly rainfall. Thus, the persistence of the crusts in the Tumas basin from the time of the gypsification of surface S1 indicates that climates have remained sufficiently dry since stage 7 to maintain the crusts. Indeed, the duricrusted surfaces represented by stages 7, 9, and 11 act as the most resistant elements in valley-bottom landscapes underlain by Namib Group sediments.

The laminar Y1m capping horizons on some gypsum crusts could well relate to the moister conditions of occasional, but more strongly flushing, rainfall. Similarly, minor iron oxide release in the Member 2 gravels accords well with local rainfall, insufficient in quantity to raise water tables. Vegetation cover while gypsum crust growth was current can never have been lush for very long.

Dating the onset of gypsification is difficult. An obvious controlling event, or environmental trend, is that of the full upwelling of the Benguela Current in late Miocene times. This change not only reduced local rainfall by generating strong atmospheric stability along the coastal Namib Desert, but also generated regular fog. The combined effect was to promote gypsum accumulation by inhibiting soil flushing by rainfall while promoting moistening of surface materials and hence gypsum mobility. The coincidental timing of the Mio-Pliocene desiccation trend inferred from the geological record seems to confirm this as a general period of gypsification of surface materials.

But the Tumas gypsums may be much older: the continued existence near surface of the Goanikontes lagoonal gypsums implies low levels of rainfall in the coastal Namib Desert at least, probably since the formation of the sediments, most likely during the Paleogene. This argument supports the mass of other geological evidence favoring an early onset of aridity in the Namib.

The notable change in chemical environment from sulfate in the Tumas Sandstone to carbonate in lower (Leeukop Conglomerate) levels could relate to the Mio-Pliocene desiccation trend: calcic horizon development in Miocene times, a period demonstrably too moist for sulfate buildup, may have given way to distal depository retreat with apparently concomitant surface sulfate accumulation.

Martin (1963) has theorized that the upper Tumas Formation strata were originally carbonate-rich, transformed latterly to gypsums under new chemical conditions of marine-derived hydrogen sulfide. No reasons were advanced to explain such a change in atmospheric behavior.

A more likely possibility is that the carbonate is a phenomenon related to groundwater, considering that the Leeukop Conglomerate underlies an existing alluvial plain and considering the likelihood of long-continued aridity evidenced by near-surface gypsum beds at Goanikontes. The carbonate/sulfate differentiation may be a syngenetic phenomenon of the kind known to operate in inland drainage basins: in fluvial systems with low gradients, brine concentrations increase both downstream and with proximity to the surface, so that sulfate facies may prograde laterally with time, overtopping the carbonate facies (Arakel and McConochie 1982:1165). In river end-point situations of vigorous evaporation, evaporitic gypsums may thus overlie "drainage" calcretes (Arakel and McConochie 1982:1167).

Stream Incision. Stages of incision 8, 10, and 12 have been ascribed to the extrinsic control of increased stream discharge under conditions of heavier local rainfall in the Tumas basin, and controls internal to the system such as adjustments of critical gradient, sediment starvation downstream of the new depository, and river flow confined by indurated stream banks. At minimum, climates producing more vigorous discharge more frequently than is the norm today must be proposed to bring the surface hydrological system of the Tumas basin back to life.

Sculpture of micropediments is conceivable only under conditions of surface flow across the weak Tumas Formation sandstones. However, implied climates cannot have been moist enough to promote wholesale karstification of the crusts.

Thus the persistence of prior gypsum crusts through phases of increased flow and localized rainfall gives apparently precise boundaries to the degree of climatic change experienced during incision phases. It must be postulated that local rainfall remained below 200 mm at maximum, or that rainfall excursions above this level were of limited duration. Without taxing the data too far, however, it can be suggested that rainfall during phases of micropediment cutting may have been confined to the narrow range between 20-50 mm of the present regime, and 200 mm, the probable upper limit of gypsum preservability.

Present and Past Morphogenesis. The alternating morphogenetic phases that appear to have characterized the lower Tumas basin landscapes have been mentioned. Arid phases were probably analogous to present-day arid conditions, and were morphogenetically stable, characterized most conspicuously by gypsum crust development, since gypsum is mobile under present near-surface moisture conditions, but cannot be flushed from the environment. Fog-derived moisture is the active agent. Another feature of this morphogenetic phase is the mobilization, also near present geomorphic surfaces, of uranium minerals, but equally, their lack of flushing from host sediments.

Fluvial activity is very restricted, and unconsolidated gypsum crust development on micropediment surfaces indicates the lack of local precipitation in the form of rainfall. Water tables are meters below surface.

The dominance of fog over rainfall precipitation thus appears to have morphogenetic consequences.

Eolian activity is responsible for such minor features as isolated coppice dunes within a few hundred meters of sand supply points in major talwegs. A small dune field at the junction of the Gawib Flats and the Tumas River (point C, figure 4) where sand supply is copious has been mentioned. Reg surfaces have developed locally by deflation of pebbly gravity flow veneers; deflation has given rise to wind streaks and two small deflation hollows in the study area.

Moister phases resurrected stream discharge along the length of the study area, with concomitant incision and elaboration of small flanking pediments cut across less consolidated Namib Group sediments. Micropedimentation processes were sufficiently energetic to remove near-stream gypsum crusts. But responsible climates were never wet enough to cause solutional removal of the crusts of prior, soil-forming, morphogenetic phases. Widespread laminar Y1m horizon development and iron oxide staining of Member 2 gravelly sands are probably phenomena that accompany moister phases. Similar polygenesis of caliche crusts is known in the Ubib area where the youngest crust has been dated at 21,280 B.P., whereas its upper laminated horizon, dated to 8385 B.P. (Blümel 1982), indicates carbonate mobilization during the moist phase that occurred between 8400 and 8200 B.P. (Sandelowsky 1977; Vogel and Visser 1981).

In their most detailed discussion of the morphogenetic phases of the Central Namib coast, Rust and Wieneke (1976) also have identified gypsum crystal growth with modern climates (their "arid stable" phase-a), with which are associated reg development and small-scale sand accumulation in the form of tied dunes. "Moist active" phases in their formulation are signified mainly by autochthonous stream incision with associated valley widening in less coherent coastal sediments. They adduce indirect evidence for increased flow in allochthonous rivers. Their third morphogenetic phase, "arid activity," consists of active dune building.

External Correlations

Phases of slope stability and duricrusting on the Tumas alluvial plain surfaces appear to be paralleled by carbonate duricrusting further inland in the Ubib River catchment and on the Tsondab Planation Surface. Surface age and percentage content of calcium compounds are correlated, suggesting progressive and probably polycyclic development in the Ubib area (Blümel 1982) and lower Tumas basin. Echoing the situation in the lower Tumas, Blümel (1982) has argued that conditions under which slope rupture occurred were insufficient to destroy the caliche crusts despite some karstic attack.

In the Ubib drainage, phases of micropedimentation have been interpreted as representing vigorous slope processes under conditions even wetter than those responsible for caliche duricrusting (Hüser 1976), on the strength of the destruction of coherent calcrete crusts and the transport of coarse rubble across the pediments. Lancaster (1984b) too has argued that incision phases in the presently dormant Tsondab drainage imply with little doubt greater-than-present discharge in the system.

Morphogenetically then, parallels with the Tumas are remarkably similar, and suggest that the three incision phases in the Tumas valley may relate to those of the Ubib drainage, and to three of the five that Marker (1979) has identified in the Tsondab drainage.

Study Area Gypcretes and Inner Namib Calcretes. Parallels are, however, far from similar in terms of the environmental demands of the two different crust types. Where sulfate crusts signify persistent aridity, carbonate crusts indicate moister-than-present climates in the Ubib (Hüser 1976) and upper Tsondab (Yaalon and Ward 1982) drainages, since these calcrete caps now exist in climates too dry for active caliche buildup (Goudie 1973) (figure 50). Yaalon and Ward (1982) have suggested an optimal annual rainfall of 400-450 mm for development of the Kamberg Calcrete, and Blümel (1982:81) "more humid conditions (up to 600 mm rainfall?)" for the development of the Ubib calcretes. Netterberg (1969) has demonstrated that in southern Africa younger elements in caliche profiles presently develop under annual precipitation regimes of 500-800 mm/yr., but warned against drawing precise environmental inferences from them.

It is possible that the Ubib and Tumas sequences are not correlated, despite striking similarities: they may have formed at different times. Whereas the morphogenetic arguments concerning intensities of operation of various processes remain, noncorrelation of the sequences means that arguments concerning rainfall gradients cannot be made. If the sequences are not coeval, the caliche crusts are probably the older, since a xeric trend would act both to preserve, though not promote development of, caliche crusts, and stimulate accumulation of gypsum crusts.

However, all the evidence suggests temporal overlap of the sequences. Blümel's (1982) Last Glacial Maximum and Holocene radiocarbon dates for the lowest calcreted surface in the Ubib area, and >45,000 B.P. date for the middle surface, give no necessary indication of the ages of similar duricrusted surfaces in the lower Tumas. But they do indicate calcrete formation into the Holocene and suggest further, therefore, that the Tumas crusts have existed coevally with the Ubib crusts, and not sequentially. Hambleton-Jones (1984) considered the Gawib calcretes at Langer Heinrich Mountain, with which the Tumas gypcretes (surface S1) are morphostratigraphically equivalent, to be precanyon in age, suggesting a Pliocene age at minimum.

It is not ascertained whether the oldest Ubib crust correlates with the Kamberg Calcrete. If it does, then the main elements of the Ubib sequence span a long period of time, from the Miocene/Pliocene boundary to the Last Glacial Maximum. The conclusion remains that sequences in the Inner and Outer Namib have developed coevally rather than sequentially.

Thus it becomes necessary to reconcile the definite rainfall increase demanded for calcrete buildup in the Inner Namib with significantly lower rainfall consistent with preservation of gypcretes in the Outer Namib. If the yearly rainfall necessary to promote caliche formation is conservatively es-

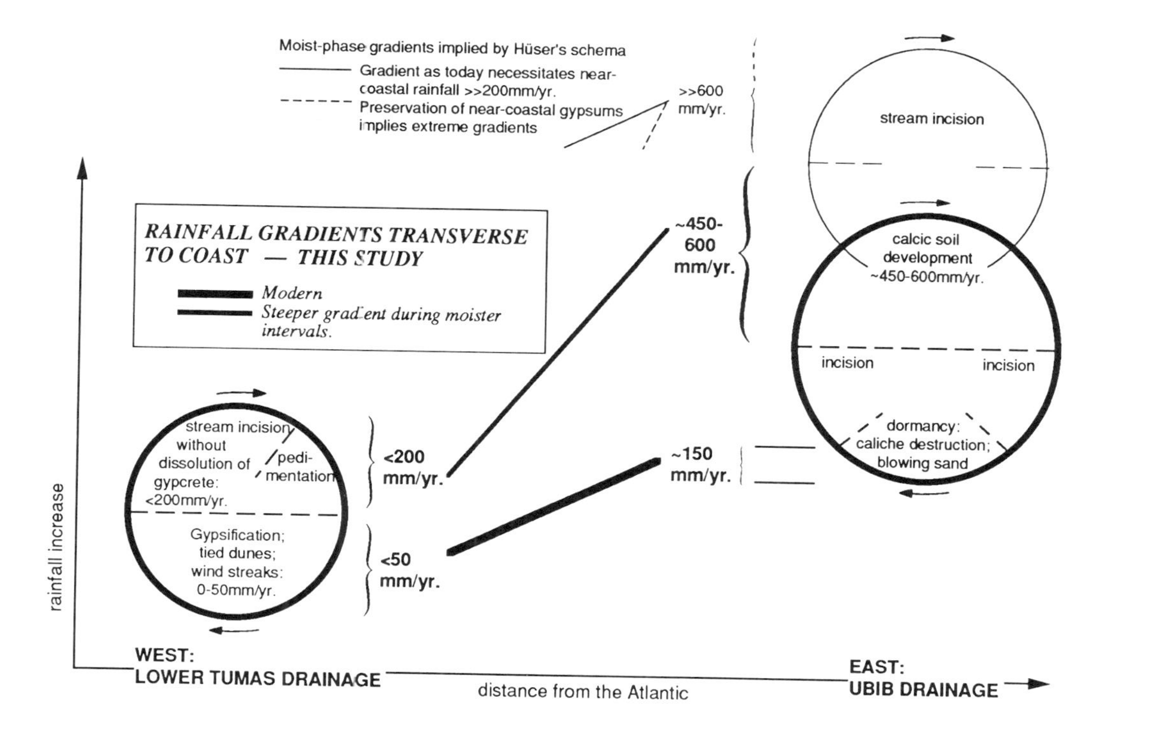

Fig. 50. Cyclic morphogenetic regimes and putative climatic relationships of major geomorphic indicators. Rainfall gradients are shown—between the lower Tumas basin regime 20-60 km inland (left) and regimes of the Ubib River basin 150 km inland (right), as reconstructed in this study (heavy circle) and by Hüser (1976) (upper circle). Moist-phase gradients implied by Hüser (1976) either necessitate coastal rainfall too high for gypsum-crust preservation or imply gradients that seem unfeasibly steep.

timated at 450 mm, then it seems that precipitation in the lower Tumas, a mere 100 to 150 km to the northwest, must have been far too high for gypsum preservation. However, since the gypsums have been preserved, a strong climatic gradient must be postulated. But even if an increase of 300 mm is imposed on the present gradient (from 150 to the requisite 450 mm in the Inner Namib and 50 to 350 mm in the study area), climates destructive of the gypsums are implied.

It is more likely that precipitation increased *relative to the present gradient,* so that a threefold annual increase yielded 450 mm in the Ubib area but only 150 mm in the lower Tumas basin. In this case, caliche formation in the Inner Namib indeed becomes compatible with gypcrete preservation in the Outer Namib. Figure 50 illustrates the notions of neighboring, though differing, morphogenetic regimes separated at different times by rainfall gradients of varying steepness.

It appears possible to propose constraints of remarkable precision, if this model holds true: specific ranges of yearly precipitation and high rainfall gradients are demanded. Although the postulated gradient is steeper under conditions of higher rainfall in the past, the effect of the Benguela Current in suppressing rainfall nearer the coast seems to be of a similar order as that of today.

Higher rainfall in the Ubib drainage, to the point of promoting caliche soil development on slopes, would presumably generate stream flow in the Namib. It is possible that such increases in discharge are linked thus to the incision phases in the lower Tumas area, suggesting correlation of morphostatic conditions in the Inner Namib (pedogenesis) with localized morphodynamic conditions in the Outer Namib (stream incision).

This correlation takes no account of the yet greater precipitation deemed necessary by Hüser (1976) for incision and pedimentation phases in the Ubib drainage. Restated, it seems that morphodynamic phases in the Ubib drainage could have no conceivable correlative in the Tumas, where the high implied precipitation would cause not only the stream to flow, but also destruction of the gypsum crusts (figure 50).

It is conceivable, on the other hand, that stream incision and expansion of micropediments in the Ubib drainage relate to more arid periods (less than 400-600 mm/yr.) (contra Hüser 1976). Hüser's (1976) arguments for climatic as opposed to tectonic initiation of incision and planation phases are fully accepted, but the precise morphogenetic mode is more difficult to ascertain. If the suggested alternative of incision and crust destruction under arid conditions held true, then stasis inland (i.e., calcreting) under wetter-than-present conditions may have corresponded with dynamic cutting and micropedimentation episodes in the lower Tumas (figure 50).

Here again, the notion of steep climatic gradients provides a solution. It is in fact the most conservative explanation which allows for concurrent development of both gypsum and calcrete crusts under virtually opposed morphogenetic regimes.

In this preferred scenario of environmental controls (figure 50), drier, morphologically active phases inland correspond with the necessarily arid phases of sulfate duricrusting nearer the coast. That wetter conditions sufficient to generate erosive stream discharge are consistent with arid climates is evident from the persistence of gypsum crusts through wetter periods in the western Tumas basin.

Ward et al. (1983) have made the same argument vis-à-vis the persistence in the landscape, probably since the late Miocene, of the Kamberg Calcrete Formation. Although conditions of calichification are far less circumscribed than those under which gypsum crusts develop, long-continued humid climates do seem to be excluded, despite reports of karstification of the Calcrete (Marker 1982; Blümel 1982; Ward 1984). Hüser's (1976) conclusion that incision phases in the Ubib represent precipitation higher than that required for caliche formation is thus at variance with the opinion of Ward et al. (1983). Both opinions may in fact be correct, however, since the Ubib data may record relatively short-term moist excursions of climate, insufficient to cause wholesale destruction of the Kamberg Calcrete. Karstic phenomena do suggest conditions wetter than those under which caliche buildup occurs.

Nevertheless, the solution suggested in figure 50 is favored here. It rests on (1) the assumption of morphogenetic cyclicity, (2) the probability of overlap in the sequences, (3) the probability of strong climatic gradients (allowing relatively high rainfall along the Great Escarpment with rapid precipitation decline westward), (4) evaluation of the evidence of karstification as comparatively insignificant on the Namib plains at the foot of the escarpment, and finally, (5) on a reinterpretation of the Ubib sequence such that alternate phases of morphodynamism are dry-related rather than moist-related.

Incision Phases. Morphogenetic arguments from terrestrial sediments along the Central Namib coast have concluded that glacially lowered sea levels correlate here with periods of higher local rainfall (Wieneke 1975; Wieneke and Rust 1975, 1976; Rust and Wieneke 1976). Since these conclusions are based in part on data that involve marine platforms of probable pre-Pleistocene age, they should be treated with caution. Further, lithostratigraphic and morphostratigraphic correlations are very difficult to ascertain in the near-coastal Tumas basin. Simple parallelism of fluvial and glacial events, although an attractive idea, has been shown to be generally false for tropical and subtropical Africa (Butzer 1974, 1976c, 1978). Present chronologies lead to the expectation that such parallelism is less, rather than more, likely to occur. The extended wetter periods required to explain incision phases in the Tumas, Ubib, and Tsondab valleys may indeed have been related to glacials, but equally they may not. Certainly it seems that the evolution of the three-phase incisional landscapes of these smaller drainages has occurred throughout the Pleistocene, vitiating support for the cyclic connection of wetter with glacial periods.

At minimum, it can be concluded with some assurance that incision phases represent greater rainfall over the Tumas watershed, because present hyperaridity represents one end of the precipitation continuum. As argued, the opposite was probably the case in the Inner Namib (figure 50). It can be concluded further that gypsum crust development represents reduced total precipitation with increased relative fog contribution. There seems less room for doubt in these interpretations than exists in the case of the calichification and incision phases in the Inner Namib, as exemplified by the Ubib landscapes.

Fluvial evidence provides clues concerning the dating of the later, erosional events in the Namib. Two major incision episodes of 30 and 80 m are documented in the Tsondab valley. The intervening terrace deposit has been equated tentatively with Middle Pleistocene silts at Narabeb (Lancaster 1984b). All these features indicate discharge greater than any at present. A prima facie equivalence of greater scouring flow seems to exist between the Tsondab, Ubib, and Tumas valley sequences. Some correlation probably does exist, especially between the larger two incision phases of the Tsondab basin and those of smaller catchments such as the Tumas and Ubib. Uranium minerals have been mobilized within the last half million years near the upper contact of the Tumas Sandstone host. Since the high water tables represented by the uranium and gley horizons probably predate incision, as has been suggested, then the incision phases (especially the later ones), may fall within the Middle and Upper Pleistocene. No certainty is possible at this stage, and incision may have begun at any time between early Pliocene and mid-Pleistocene times.

Present attempts at organizing the Tertiary history of the Namib concern putative correlations of similar but disjunct lithological and geomorphic successions. The Tumas history appears to fit fairly well the regional aggradational sequence as outlined by Martin (1950, 1961) and Ward et al. (1983), and local erosional sequences described by Hüser (1976), Blümel (1982), and Lancaster (1984b). However, the method is a first approximation only and awaits improvement by contributions of further, detailed local studies and improved absolute dating. The picture is likely to be more complex than the present regional sequence suggests, indispensable as the preliminary chronology remains. Sequences in different basins are likely to diverge: thus, although the younger deposits of several of the exoreic streams of the Namib include Homeb-like silts (Ward et al. 1983), the Swakop River does not. The complexities of hinterland climate are reflected in the more complex late Cenozoic chronologies of exoreic river deposits; the endoreic Tumas, Tsondab, and Ubib rivers, on the other hand, show broad similarities and lesser complexity, while otherwise conforming to the regional pattern.

The latter, small drainages may ultimately give a more accurate picture of environmental fluctuations in the Namib per se, uncomplicated by inputs generated on the plateau that affect the larger rivers. Even in the case of a desert-bound stream such as the Tumas, the phenomenon of climatic gradient seems to have been a persistent theme in the past.

These observations may explain the several, well-documented late Pleistocene (and Holocene?) phases of increased carbonate mobilization in the Outer Namib along the Kuiseb River tract, southward in the dunes and at coastal sites. All these sites lie within drainage systems (subsurface or otherwise) that rise on the escarpment or further inland. These topographic locations may explain the existence of high water tables and carbonate close to the coast when gypsums are patently pristine and show no sign of removal or attack, immediately north of the Kuiseb River and beyond. It has been shown that conditions of carbonate mobilization are antagonistic to sulfate crust preservation. Gypcreted sediments typically lie within small drainage basins (notably unaffected by large drainage systems), which rise west of the escarpment.

In short, it seems necessary to propose that the wet phases—so dramatically and consistently documented by numerous radiocarbon dates—may reflect precipitation increases further inland near, on or beyond the escarpment. In the case of zones in the ambit of the Kuiseb River, where various changes in the courses of the Kuiseb and Tsondab drainages are well documented (Goudie 1972; Marker 1977; Rust and Wieneke 1974; Lancaster 1984b), the dates may reflect conditions tens or even hundreds of kilometers inland.

These conclusions stand even in the face of such complicating factors as changing precipitation gradients, changing hydroclimates offshore, and changes in windiness and rainfall seasonality.

Young Features: Late Pleistocene and Holocene

Numerous submodern features exist on the Tumas Flats. These marks of hyperaridity provide a yardstick against which earlier landscapes may be evaluated. Of all these features, wind streaks have proved particularly useful in giving chronologic perspective to a major phenomenon of Namib geohistory, namely the development of the dominant north-south seif dunes of the Namib sand sea.

The weight of evidence favors a diachronic model of evolution for the dune sea south of the Tumas basin. This means that eolian features in the study area (wind streak patterns and ventifaction especially) are submodern to Holocene, whereas dune patterns in the sand sea to the south, except for superficial remodeling, are best regarded as fossil features (inland of the present narrow belt of highly active coastal dunes). The diachronic argument postulates that southerly winds, more like those that blow along the coast today, were indeed formative of the large linear dunes that make up the dominant dune pattern in the Namib sand sea. The atmospheric circulation responsible for such winds was arguably one of a translocated or more vigorous South Atlantic anticyclone, probably during glacial episodes. This conclusion comes from evidence for contemporary—and probable Holocene—activity of northeasterly "Berg Winds," winds that have since obliterated any vestige of prior southerlies in the Tumas basin. The northeasterly winds are apparently remodeling the Pleistocene face of the dune sea, in its northern half at least.

The long-continued action of the Berg Winds in the lower Tumas basin suggests that conditions have been substantially like those of today for the duration of the Holocene, with formative winds from the northeast (under conditions of aridity)—unlike hypothermal periods in the Pleistocene when general windiness was probably greater, with southerly winds on average stronger and geomorphic effects of the Berg Wind presumably reduced in relative terms.

The late Cenozoic has been conducive to dune sand accumulation in parts of the Namib characterized by decreasing wind speeds, as in the area of the present dune sea (Lancaster 1985). Ward et al. (1983) have suggested that the Sossus Formation and Obib dunes are Pliocene in age. Besler (1980, 1984) has noted that the volume of material excavated from the incised valleys in the dune Namib is crudely equivalent to the volume of dune sand west of the river end-point playas. This observation implies that the dunes may postdate the first major fluvial incision phase that released the sand for eolian sculpture.

Morphogenetic equivalents of Plio-Pleistocene sand accumulation are pedogenic sulfate crust development and reg development by deflation.

Chapter 10

CONCLUSIONS

I wished to go in a whale boat to the mouth of the Swakop, to ascertain the existence of elephants, which are said to be numerous about the mouth.

The effect of the late rains began soon to show itself, for even the barren Naarip was in places richly carpeted with grass and flowers. . . . Herds of the beautiful oryx, the lively quagga, and the grotesque gnoo . . . served further to enhance . . . the scene. These were glorious times for the lions, who were exceedingly numerous.

—Sir J. E. Alexander, *An Expedition of Discovery into the Interior of Africa*

And no man will ever again see a herd of four thousand springbok in this neighbourhood as once we did.

—Henno Martin, *The Sheltering Desert*

Indicators of the earliest paleoenvironmental conditions preserved in the sedimentary and geomorphic record of the lower Tumas basin are equivocal, but later indicators are more illuminating. These indicators are of five main types: (1) deposits of clearly definable geomorphic and paleoenvironmental setting, (2) behavior of the sediment dispersal system, (3) soil and groundwater phenomena, (4) a long record of stream behavior with striking parallels in neighboring drainages, and (5) some of the more persuasive data concerning the polyphase nature of formative wind regimes in the Central Namib.

This evidence has given a long history of environmental change in the Central Namib, changes with both long and short periodicities. The later changes are more detailed in terms of periodicity and degree of paleoenvironmental change documented.

Chronologically, the changes identified in this study are (1) an early phase of aridity, probably of long duration, beginning in the Paleogene. Oscillations to the moist side probably occurred, and may have been as moist as the next phase: (2) a pan-Namib phase of semiaridity during the Miocene documented by increased river discharge and massive, calcic soil development, followed by (3) a progressive reversion to aridity, with evidence for establishment of the present, steep climatic gradient, culminating in accumulation of eolian sands of the present dune sea.

Two variants on the last are conditions of even greater aridity and stronger winds, which may have been required to generate the linear dune chains of the Inner Namib. The other variant encompasses episodically moister phases, some generated inland and probably fewer indicating rainfall increases within the Namib Desert itself.

Namib Group sediments host above-background concentrations of uranium, which first attracted attention in the 1970s. The disposition of these deposits has allowed insights into the nature of the Namib Unconformity Surface (NUS) on which they lie, into their architecture, and into patterns of paleodrainage. The disposition of the older sediments was probably controlled broadly by long-term eustatic movements of sea level.

Planed and incised surfaces cut across ancient rocks separate the sedimentary bodies in and around the study area, surfaces regarded as part of the NUS of Ollier (1977). The NUS proves to be a more complex, multiphase feature than generally represented. Where planed in the study area it appears to be a comparatively young feature. In the southern Namib it seems to be a simpler, older feature.

Sedimentation levels on the NUS were likely determined by global sea level. Consideration of such altitudes may explain one of the early events in the Tumas sequence, namely the reorientation of the lower courses of the Swakop and Khan rivers. Sedimentation levels may have breached divides in the lower course of the proto-Swakop River, thereby facilitating the river's reorientation from an original course occupying the Gawib Flats and lower sectors of the Tumas basin—a kind of self-induced piracy under conditions of river aggradation rather than erosion. Reorientation led to disconnection of the Tumas drainage from the Atlantic Ocean, with resulting endoreic geomorphic features and styles of sedimentation. The lower courses of other west coast rivers (Kuiseb, Orange, and Olifants) have undergone reorientation, suggesting that this is not an abnormal phenomenon on trailing coasts subjected to a series of regressions and transgressions.

Final planation of ancient, relatively hard rocks was substantially completed at present levels in relation to the final major phase of sedimentation during the mid-Tertiary. It has been suggested that the number of post-NUS sediment bodies in and around the study area is numerous enough to justify the use of the umbrella name "Namib Group."

Paleoenvironments and Their Sequence in the Central Namib

A major conclusion of this study, one that provides a background to the rest, is that the extant indicators, as well as implied formative environments, all point directly, some more specifically than others, to arid and semiarid climates throughout the history of the Tumas basin. Sediments in the catchment may date back as far as the early Eocene and possibly to terminal Cretaceous times.

This conclusion accords with those of the majority of studies of sediments in the Namib, and supports the arguments of Ward et al. (1983). These researchers have questioned conclusions based on mainly biological data. Where the marine and paleontological evidence suggested that Paleogene terrestrial environments were mesic in the Namib, sedimentological and other evidence has suggested that most of the Tertiary was arid or semiarid. The tenor of conclusions derived from these bodies of data differs in this essential. Evidence from the present study consistently concurs with other sedimentological evidence favoring the dominance of aridity or semiaridity for most of the Tertiary at least. This issue is discussed at greater length below.

Evaporitic deposits, represented by the Goanikontes formation, although disjunct from the axial body of sediments in the lower Tumas basin, are probably the earliest indication of aridity in the coastal Central Namib. The deposit is the coastal extremity of voluminous valley-fill sediments of the younger Namib Group rocks. These deposits were laid down by the large drainages of the area (Swakop, Khan, and Kuiseb), all of which converged, then as now, on the Swakopmund–Walvis Bay embayment of the coastline. They comprise several large, discrete bodies, defined by mineral exploration activity, which swell the number of known formations in the Namib Group.

The largest deposit west of the escarpment to provide data on mid-Tertiary aridity is the Tsondab Sandstone Formation, with which the upper units of the Tumas Sandstone Formation appear to be correlated. The existence and location of the former, significant eolian body implies certain long-term climatic controls of specific interest in reconstruction of a history of the Namib Desert. The Tsondab Formation is briefly discussed here as a context within which to view evolution of the Tumas Sandstone.

For various reasons, but especially climatic reasons, the general region of the Namib Desert occupied by the Tsondab Sandstone appears to have acted for a long time as continental sediment sink. Besler (1984) has suggested that the main source for the sandstone has been the small, local rivers that end in the dune sea. Because of both their small size and the aridity of the climate, these fail to reach the ocean, their loads deposited on land as a readily deflatable source of eolian deposits. Lancaster and Ollier

(1983) have proposed that blown sands from southerly source areas tend to accumulate in the dune sea since southerly winds experience significant reduction in velocity in the region between the south and north ends of the dune sea.

There is no reason to expect that the Tumas basin was not also part of this great sediment sink on the most arid parts of the west coast.

Since permanent oceanic upwelling was established in early Upper Miocene times, the accumulation of dune sand represented by the Tsondab Sandstone in preupwelling times—possibly for 20-30 million years before the late Miocene moist episode (Ward et al. 1983)—is an event that focuses on the fact that upwelling per se is not a prerequisite for the accumulation of fluvial and eolian deposits in endoreic basins. Upwelling does not accompany the majority of the earth's dune fields; just as the Benguela Current is probably insignificant in understanding the development of present or past dune fields in the Kalahari Desert, it need not be invoked as a necessary control in the origin of arid-related deposits on the west coast.

The existence of a dune field in the Tumas basin corresponds with the regional geological sequence precisely. It fits the pattern of dune accumulation in the other endoreic catchments of the Namib such as the Ubib, Tsondab, Tsauchab, and others further south. It also accords with the fact that the Tsondab Sandstone occupied a larger area than it does today, after extensive peripheral erosion (Ward 1988). Other smaller dune fields exist to the north of the Tumas basin.

By contrast with the apparently long dry period in the early- to mid-Tertiary, the fluvial emplacement of Members 1 and 2 of the Tumas Sandstone Formation represents a period of greater rainfall. Evidence of fluvial reworking of dune sand is so widespread that a similar event in the Tumas basin seems likely to have been contemporaneous. That the period was probably long continued is shown by the thoroughness of the resculpture of dune-field surfaces by water, into the form of pediments in the catchments south of the Kuiseb River (the Tsondab Planation Surface of Ollier 1977), and, if interpretation of a prior Tumas dune field is correct, by the complete reworking of dune sand so that eolian sedimentary structures are nowhere evident today.

Parallels with the Tsondab Sandstone are apparent. Ward (1988) has recently concluded that the sandstone comprises not only an eolian facies, but related fluvial and playa facies indicative of arid environments. Whereas the Tumas River was capable of reworking a possibly smaller dune field, less fluvial reworking of the dune component was achieved in the case of the great sand field of the Tsondab Sandstone.

The degree of rainfall increase is less easy to gauge in the southern catchments, but sediment characteristics in the Tumas basin, both diamictic

Sm lithofacies and the sheetlike architecture of Member 1 (and 2), all point to terminal aggradation on an alluvial plain dominated by mass flow and sheet-flood deposits, a zone not necessarily coincident with the topographic low point of the basin.

Thus it may be concluded that (1) an increase in rainfall gave rise to fluvial deposition where such activity had not existed before, and (2) that the data permits refinement of this conclusion: the increase was not sufficient to allow the Tumas River access to the sea. At most, a semiarid climate is implied, probably only along the Great Escarpment.

Evidence for alluvial deposits, extensions of now truncated rivers, and better integrated drainage patterns all support the notion of greater moisture in the Namib Desert during this period. In the Central Namib, extension of the lower Tsondab drainage system and its connection with the Kuiseb canyon (in the vicinity of Gobabeb) is proved from the distribution of fluvial gravels on the infradune Tsondab Planation Surface (Lancaster 1984b).

The meaning of the change in chemical environment from carbonate in the Leeukop Conglomerate Formation to sulfate in the Tumas Sandstone is unclear. Martin (1963) has suggested that the upper Tumas Formation strata were originally carbonate-rich and have been transformed progressively to gypsums under new chemical conditions of marine-derived hydrogen sulfide. No reasons were advanced to explain the change in atmospheric behavior. However, it is concluded that the change most probably reflects a syngenetic evolution of sulfate facies overtopping the carbonate facies in the end-point setting of an arid river.

Evidence for high water tables in the Tumas Formation is abundant but poorly dated. The highest undoubtedly predate the first stage of incision of the Tumas plain. The detail of various exposures shows hydromorphic and uranium-precipitate evidence of rising, falling, and permanently high water tables, probably on several occasions. The phenomenon of high water tables seems to fit the local and regional chronology as part of the fluvial period when more rain fell in the smaller Namib catchments. High water tables are not incompatible with proven xeric conditions in the lower Tumas basin, because of the strong rainfall gradients induced by (1) distance from cold hydroclimates at the coast and (2) orographic effects of the escarpment. This problem receives further attention below.

The present gradient climbs steeply from ca. 25 mm/yr. in the coastal tract to 150 mm/yr. in the Inner Namib 140 km inland. Anything like a similar gradient imposed on generally higher levels of precipitation will have resulted in significant rainfall inland, thereby raising water tables along drainage lines in the more arid west. High rainfall and gypsification are not compatible. Steep rainfall gradients, with concomitant allochthonous, escarpment-derived water along drainage lines in the hyperarid core, are required to explain their coexistence in the western Tumas drainage.

The third major climatic regime is the trend toward aridity. The phase of modest increase in rainfall, implied by emplacement of Tumas Formation Members 1 and 2 and high water tables, was followed by the retreat inland of the Tumas sediment depository. Recession of the locus of deposition, by 10-20 km at minimum, in a small drainage, implies distinct aridification. This event marks a reversion to conditions conducive to the evolution of sand seas as represented by the Tsondab Sandstone and the present sand sea. It is also represented in the geomorphic sequence by gypsum crust development in the study area.

It is remarkable that the existence of surface gypsums, long known, has received so little comment in the evaluation of Namib paleoenvironments. The first significance of the gypsum crusts, because of their innate solubility, is that they develop only in the earth's driest climates where soil flushing is nonexistent. Second, in the particular situation of the Namib, the persistence of gypsum crusts through subsequent events of manifestly moister character allows unexpected refinement in understanding the amplitude of the moist climatic excursions. Though highly prone to karstic attack, the persistence of duricrusts, the most resistant features of the younger landscapes, is strong evidence that rainfall of more than approximately 200 mm/yr. did not occur in the Outer Namib for extended periods. Rainfall may have reached twice or three times this figure inland, and even more for short periods, under operation of the controls that determined the acute climatic gradient.

Similar arguments have been applied to the presence in the Inner Namib of the massive Kamberg Calcrete (Yaalon and Ward 1982), a unit possibly of late Miocene age. In the case of a carbonate duricrust, the parameters implied by its preservation are very much wider: Netterberg (1969) documented extant, though probably decaying, calcretes in northern Botswana under present annual rainfalls of more than 650 mm/yr. Karstification of the widespread Inner Namib calcretes has been mentioned, and whereas Marker's (1982) view that karstification belies wetter climates is accepted, such climates appear to represent lesser oscillations in the sweep of past environments, sufficiently short-lived or infrequent to prevent the destruction of duricrusts.

The calcreted landscapes of the Inner Namib and the gypcreted landscapes of the Outer have in all likelihood developed pari passu for a long period of time. This conclusion raises problems concerning the coexistence of morphogenetic systems with mutually exclusive pedogenic requirements. I have concluded (1) that calcic soil development inland best parallels periods of stream incision in the arid Outer Namib, without the destruction of gypsic crusts; and (2) that periods of incision inland, although ascribed to much wetter climates by Hüser (1976), are unlikely in terms of feasible rainfall gradients or gypsum preservation. I have argued that incision may have occurred as easily under climates drier than those that gave rise to calcification. Such conditions then correlate well with patently morphostatic gypcreting environments in the more arid west.

Much evidence exists for the moist oscillations of climate within the broadly arid post-Miocene period. It is against the background of aridity that phases of stream incision in the Tumas and other catchments took place. I have argued that rainfall higher than present must be invoked to explain 20 m of incision—and associated micropedimentation—in the Tumas drainage. Lancaster (1984b) has argued similarly for incision phases, totalling 110 m, in the larger Tsondab catchment, which rises further east in the semiarid Naukluft Escarpment. If these cyclic stages represent periods as wet as the Miocene fluvial phase, which seems unlikely, then they imply short periods of such activity. Incision and micropedimention are compatible with periods of moderate increases in rainfall as evidenced by the fact that duricrusts have outlasted the wetter phases with which incision phases are associated.

The more complex history of the Kuiseb River basin and semiallogenic groundwater flow in the Tsondab and Tsauchab basins is undoubtedly climatically controlled and shows a series of moister climatic excursions, against the background of Plio-Pleistocene aridification. The most prominent and well-defined excursions are an early phase terminating at or before the Last Glacial Maximum and another coinciding with northern hemisphere deglaciation (14,000-8000 B.P.).

Vogel (1982) distinguished between Namib and hinterland precipitation and suggested that periods of higher rainfall in the Outer Namib persisted until ca. 28,000 B.P. Three moister-than-present Holocene phases are indicated from evidence in rock shelters in the Outer Namib and likely relate to autochthonous rainfall. Many radiocarbon dates come from materials situated in basins draining wetter areas to the east. The latter dates may thus reflect more on rainfall conditions inland than humidity in the Namib itself. The importance of the climatic gradient is such that geomorphic setting of dated materials must be taken into account when evaluating clusters of radiocarbon dates. Date clusters in the paleo-drainage of the Kuiseb River have been taken to mean increased Pleniglacial or late glacial moisture, but may reflect conditions in the hinterland beyond the Namib Desert. This history of allochthonous flow includes radiocarbon data from Meob and Conception bays. These coastal localities are the distal zones of allogenic drainages that rise above the escarpment. Dates reflecting soil moisture may thus represent conditions in the Outer Namib, the escarpment, or both.

Whereas allogenic stream flow probably reflects large-scale atmospheric circulation over the subcontinent, autogenic flow probably concerns relative intensities of and interactions between large-scale circulations and local atmospheric effects related to upwelling.

Paleoclimatic Models and Controls

Implications for paleoenvironmental change documented in the Tumas and neighboring endoreic drainages are severalfold. The first is that a slave relationship does not necessarily exist between Namib climate and the existence, intensity, and position of the zone of Benguela Current upwelling. The earliest major paleoclimate identified in this study is a possibly long period—Ward et al. (1983) have suggested that it may have endured for at least half the Tertiary—of eolian sand accumulation, well attested by the Tsondab Sandstone and implied by textural characteristics of the Tumas Sandstone.

A paleoclimatic model sufficient to account for this early, arid, and undoubtedly lengthy period of time is at variance with an upwelling-related model, since the eolian period probably dates back to times prior to permanent upwelling in the Benguela Current (pre-late Miocene). Further, the eolian period is one during which marine, faunal, and terrestrial floral indicators paradoxically suggest mesic environments, the more remote in the time, the more humid.

Ward et al. (1983) suggested a solution. The paleoenvironmental meaning of the biological evidence may be complicated in unexpected ways by the narrowness of the Namib Desert and by the poorly understood behavior of desert fauna, behavior that may include migrations into the most arid core areas.

The second major climatic trend identified in this study was probably confined largely to the escarpment zone and the plateau beyond. The model of aridification induced by increasing upwelling and declining ocean temperatures fails to explain the Miocene fluvial event that occurred as upwelling became less spasmodic and more permanent. By the upwelling model, enhanced aridity should be the climatic expression of hydroclimatic temperature decline. Furthermore, since the axis of aridity is located along the coastline, eustatic recovery from very low Miocene levels will have shifted both coast and local aridity progressively east as far as the lower Tumas basin. For these reasons a pan-Namib phase of greater stream activity in the Miocene seems out of kilter.

Although the fluvial phase is probably not even subhumid in character, the subsequent (third) climatic shift is one of detectable drying marked by degeneration of the Tumas and other smaller drainage systems. The trend toward active sand transport by wind accords precisely with progressive aridification, which is documented in so much greater detail for the

Pliocene and Pleistocene. Here the terrestrial and hydroclimatic environments appear to march together.

These climatic changes demand an explanation. Ideas can be grouped as local effects, especially coastline shifts, and the synoptic effects of cloud band evolution.

Climatic effects of the Benguela Current are by no means well understood. Ward et al. (1983) have drawn attention to the zonal belt of atmospheric subsidence as a potent aridifying agency, which exists quite apart from the effect of the Benguela Current. At local scales Besler (1972) and others have noted that modern, near-surface effects of the current on the atmosphere are limited to a coastal zone only 40-60 km wide. If it is indeed the case that direct effects of the current are not felt beyond the coastal tract of intense, low-altitude stability, then westward migration of presently marked climatic gradients imposed by coastline shifts may explain various paleoenvironmental changes in the Namib. Certainly a westerly shift of rainfall gradient during periods of lowered sea level, combined with local effects of warmer offshore water, for example, could significantly increase precipitation in a near-coastal tract such as the western Tumas basin. Attenuation of the gradient without a higher sea-surface temperature might achieve as much.

At a synoptic scale, the subcontinental effects of the Benguela Current may be other than usually modeled. Tropical disturbances may be forced by its presence to become stationary over the subcontinent, typically over northern Botswana, southeast Angola, and northern Namibia, in their present fashion (Harrison 1988). Were the cold water offshore to disappear, as it has been known to do at times in the past, tropical disturbances might continue migrating westward, thereby reducing precipitation in these areas.

This argument can be taken further. Tropical disturbances act as the anchor point for bands of cloud that trend southeast from southern Angola–northern Namibia. These "cloud bands" link tropical systems with passing westerly waves and are an expression of the dominant rain-bearing systems over southern Africa today (Harrison 1984, 1986). The existence of an upwelling Benguela Current may thus exert feedback effects enhancing cloud band formation and increasing rainfall over much of the subcontinent. This is the reverse of the usually understood mechanism. Whereas upwelling is undoubtedly responsible for coastal and near-coastal aridity, its effect on wider climatic patterns is by no means established and may indeed have caused an increase in subcontinental rainfall.

These considerations may help explain the conundrum of dune-field evolution in preupwelling times. Although the existence of Antarctic ice and cold upwelled water has been seen as part of a comprehensive explanation for the origin of the Namib Desert, present zonal circulations, long continued, may be sufficient explanation for the existence of desert and dunes. Part of such an explanation may be lower precipitation induced precisely by the lack of upwelling, by consequent unchecked westerly migration of tropical systems, and by the failure of cloud band development.

It is not my intention to discuss the general circulation models (GCMs) for various paleoenvironments. Opposing models, of penetration into the Central Namib by temperate versus tropical systems, have been presented (chapter 4). In common with other parts of the subcontinent, too little data is yet available for the Namib to allow solution of this problem. It is sufficient to note that recent cloud band research (Harrison 1984, 1986) has stressed the importance of a tropical easterly wave component. Since tropical systems are the most important single element in the scheme, it seems at present that these systems may be the most important determinants of Namib paleoclimates, as Butzer et al. (1978) proposed. The location of the Central Namib well within the ambit of present tropical regimes supports this contention. Furthermore, newly proposed teleconnections, via the Southern Oscillation, between the supply of moisture to southern African air masses and a source in the western Pacific (Lindesay et al. 1986) could prove particularly important.

Terrestrial biological evidence has generally been interpreted as supporting the model of climatic change closely associated with the timing, position and intensity of upwelling in the Benguela Current. The paucity of evidence on which this model is based is apparent from the fact that different authorities have proposed Oligocene (van Zinderen Bakker 1975), late Miocene (Siesser 1978), Plio-Pleistocene (Tankard and Rogers 1978), and even early Pleistocene (Axelrod and Raven 1978) dates for the onset of aridity. Despite ever firmer dating of the stages in establishment of upwelling, it is argued here that upwelling may have less to do with the origin of the Namib desert than has been thought.

The fluvial wet phase is not accounted for in either the sedimentological model of Tertiary aridity or the biological model of progressive Tertiary aridification. Increased moisture on the Khomas Hochland and plateau, and associated increases in the flow of smaller Namib rivers draining the escarpment, may have been associated with the onset of upwelling and the consequent positioning of cloud bands over the subcontinent. Locally, upwelling probably generated progressive desiccation that ultimately counteracted the trend to higher rainfall within the confines of the Namib Desert itself as the coastline shifted east and approached its modern location. It is suggested that the present climatic gradient may have evolved in this manner.

By this scenario, flow in the larger Namib rivers can be reconciled with upwelling, and should indeed be regarded as an appropriate trend.

This construction raises problems in understanding the progressive drying of the plateau, the third, major climatic trend. Three explanations suggest themselves. (1) With time, upwelling has indeed increased above that characteristic of the Upper Miocene, thereby extending its effect inland. (2) The dynamics of tropical systems may have changed in response to other controls, especially teleconnections with the western Pacific. (3) Enhanced Pleistocene windiness may have increased eolian effects by enhancing evaporation, an effect postulated for arid Australia (Bowler and Wasson 1984) during the Last Glacial.

The fourth paleoclimate may be considered a subset of the third, namely moist oscillations away from long-term aridity. A number of scattered and as yet ill-correlated episodes of wetter climate in the arid Namib are known. Causes for fluctuations may relate to controls (1), (2), and (3) above working in the opposite direction.

A fifth climatic regime, representing those oscillations that appear to have been responsible for the construction of the major, linear dune masses of the Namib Sand Sea, may have been characterized by an intensification of controls (1), (2), and (3). It has been concluded that these features are not forming now, that they may relate to increased wind velocities of glacial times, and that they are presently undergoing a process of refashioning under the influence of northeast winds that are today the strongest sand-moving winds in the Namib Desert except for the narrow coastal tract.

Dominant, present-day eolian features in the study area are wind streaks, demonstrably the products of northeastly winds. The study area has apparently experienced stronger southerly winds than exist today, winds that may have rivaled present northeast winds in speed and geomorphic importance. If the relative size of dunes associated with these winds is any indication, then southerly winds have been dominant in the Central Namib, arguably for longer spells of the Pleistocene than have present wind patterns.

Evidence accumulates that many younger bodies of sediment exist in the Namib in what has been termed here the Namib Group of sediments. Confirmation, or otherwise, of the broad correlation of geological sequences in other basins—work that has begun in the Kaokoveld drainages of the northern Namib (see, e.g., Rust 1987, 1989; Rust and Vogel 1988)—is an important task for future research. As important is continued investigation of the disjunction between the faunal and sedimentological evidence of past climates.

Investigation of the Tumas sediments leads to the conclusion that at least some sedimentary environments are susceptible to analysis of climatic controls via architecture of the sediment bodies. For this reason, analysis of architecture is seen as an additional tool (and a formalization of much-practiced aspects of fluvial sedimentology) in adducing paleoenvironments from fluvial sediments. Closer study of geomorphic environments of sedimentation, and hence of the effect of morphogenesis on fluvial architecture, is likely to be a fruitful realm of research as databases allow reconstruction of three-dimensional configurations of sediment bodies, and as fluvial subenvironments and associated sediment morphology become better known.

An attempt was made in chapter 5 to investigate the connection between attributes of the Tumas sediments and the confined arid alluvial plain, a sedimentary environment that has not been modeled heretofore. Such attributes as sheetlike architecture, dominance of gravity flows, little or nonexistent channeling of flow, confining valley walls, convexity of bounding surface at different scales, terminal sedimentation on a floodplain, and the lack of playa facies characterize the model.

Palaeoclimates of the Namib Desert appear to have been dominated by an interplay of synoptic-scale winds and the effects, local and possibly subcontinental in scale, of the Benguela Current. The interplay has generated an east-west climatic gradient that has undoubtedly fluctuated through time, and that complicates the evaluation of sedimentological, faunal, and floral data. Climatic requirements for different soils can be reconciled only by invoking changes in past climatic gradients. Climatic explanations for winds responsible for the existence of the great linear dunes, if they are indeed paleoforms rather than modern, implies changes in atmospheric circulations.

The evolution of the Namib Desert is an intriguing, many-faceted question in which a growing number of disciplines are showing interest. Understanding past environments of the Namib Desert will rely not only on the collection and integration of a variety of data from many other Namib drainage basins. It will rely also on improved understanding of the present dynamics of southern Africa's atmospheric circulation, the study of which is presently in a state of flux. The direction in which such studies proceed will influence the way field data is evaluated. We can look forward to fruitful advances as results from field data interact with new climatic theory.

Bibliography

Ager, D. V. 1980. The nature of the stratigraphical record. 2d ed. New York, John Wiley. 122 pp.

Alexander, J. E. 1838. An expedition of discovery into the interior of Africa. London, Henry Colburn, vol. 2. 306 pp.

Allanson, B. R. 1984. The symposium in retrospect. Symposium on the Benguela "warm event" of 1982-3. South African Journal of Science 80:50-51.

Allen, J. R. L. 1965. A review of the origin and characteristics of recent alluvial sediments. Sedimentology 5:91-191.

_____. 1974. Studies in fluviatile sedimentation: Implications of pedogenic carbonate units, Lower Old Red Sandstone, Anglo-Welsh outcrop. Geological Journal 9:181-208.

_____. 1978. Studies in fluviatile sedimentation: An exploratory quantitative model for the architecture of avulsion-controlled alluvial suites. Sedimentary Geology 21:129-147.

_____. 1983. Studies in fluviatile sedimentation: Bars, bar-complexes and sandstone sheets (low-sinuosity braided streams) in the Brownstones (L. Devonian), Welsh Borders. Sedimentary Geology 33:237-293.

Andersson, C. J. 1856. Lake Ngami; or explorations and discoveries during four years' wanderings in the wilds of southwestern Africa. London, Hurst and Blackett, 2d ed. 546 pp.

_____. 1861. The Okavango River: A narrative of travel, exploration, and adventure. London, Hurst and Blackett. 364 pp.

_____. 1875. Notes on travel in South Africa. London, Hurst and Blackett. 338 pp.

Andrews, W. R., and Hutchings, L. 1980. Upwelling in the southern Benguela Current. Progress in Oceanography 9:1-81.

Angell, J. K.; Pack, D. H.; and Dickson, C. R. 1968. A Lagrangian study of helical circulations in the planetary boundary layer. Journal of Atmospheric Science 25:707-717.

Arakel, A. V., and McConochie, D. 1982. Classification and genesis of calcrete and gypsite lithofacies in paleodrainage systems of inland Australia and their relationship to carnotite mineralization. Journal of Sedimentary Petrology 52:1149-1170.

Axelrod, D. I., and Raven, P. H. 1978. Late Cretaceous and Tertiary vegetation history. In M. J. A. Werger (ed.), Biogeography and ecology of southern Africa, pp. 77-130 (Monographiae Biologicae, vol. 31, pt. 1). The Hague, W. Junk.

Bagnold, R. A. 1953. The surface movement of blown sand in relation to meteorology. Special Publication, Research Council of Israel, no. 2, pp. 89-93.

Baines, T. 1864. Explorations in South-West Africa. London, Longman.

Baker, V. R. 1978. Adjustment of fluvial systems to climate and source terrain in tropical and subtropical environments. In A. D. Miall (ed.), Fluvial sedimentology. Memoir, Canadian Society of Petroleum Geologists, no. 5, pp. 211-230.

_____. 1986. Fluvial landforms. In N.M. Short and R.W.Blair, Jr. (eds.), Geomorphology from space: A global overview of regional landforms, pp. 255-316. Washington, D.C., National Aeronautics and Space Administration.

Baker V. R.; Kochel, R. C.; Patton, P. C.; and Pickup, G. 1983. Palaeohydrological analysis of Holocene flood slack-water sediments. In J. D. Collinson and J. Lewin (eds.), Modern and ancient fluvial systems. Special Publication, International Association of Sedimentologists, no. 6, pp. 229-239.

Baker V. R.; Pickup, G.; and Polach, H. A. 1983. Desert palaeofloods in central Australia. Nature 301:502-504.

_____. 1985. Radiocarbon dating of flood events, Katherine Gorge, Northern Territory, Australia. Geology 13:344-347.

Bang, N. D . 1973. The southern Benguela system: A finer oceanic structure and atmospheric determinants. Ph.D. thesis, University of Cape Town.

Barnard, W. S. 1964-65. 'n Kaart van die Klimaatstreke van Suidwest-Afrika. Journal of the South West Africa Scientific Society 18-19:74-84.

_____. 1973. Duinformasies in die sentrale Namib. Tegnikon (Pretoria), December, pp. 2-13.

Beaudet, G., and Michel, P. 1978. Recherches géomorphologiques en Namibie Centrale. Recherches Géographiques à Strasbourg (special no.). 139 pp.

Beaumont, P. B., and Vogel, J. C. 1972. On a new radiocarbon chronology for Africa south of the Equator. African Studies 30:155-182.

Beerbower, J. R. 1964. Cyclothems and cyclic depositional mechanisms in alluvial plain sedimentation. Kansas Geological Survey Bulletin 169:31-42.

Beetz, W. 1926. Tertiärablagerungen der Küstennamib. In E. Kaiser (ed.), Die Diamantenwüste Südwestafrikas (Berlin), Dietrich Reimer, 2:1-54.

Bellair, P. 1954. Sur l'origine des depôts de sulphate de calcium actuels et anciens. Academie des Sciences (Paris), Comptes Rendus 239:1059-1061.

Besler, H. 1972. Klimaverhältnisse und klimageomorphologische Zonierung der zentralen Namib (Südwestafrika). Stuttgarter Geographische Studien 83. 209 pp.

_____. 1975. Messungen zur Mobilitäte von Dünensanden am Nordrand der Dünen-Namib (Südwestafrika). Würzburger Geographische Arbeiten 43:135-147.

_____. 1976. Wassüberformte Dünen als Glied in der Landschaftsgenese der Namib. Mitteilungen, Basler Afrika Bibliographien 15:83-106.

_____. 1977. Untersuchungen in der Dünen-Namib (Südwestafrika). Journal of the South West Africa Scientific Society 31:33-64.

_____. 1980. Die Dünen-Namib: Enstehung und Dynamik eines Ergs. Stuttgarter Geographische Studien 96. 241 pp.

_____. 1984. The development of the Namib dune field according to sedimentological and geomorphological evidence. In J. C. Vogel (ed.), Late Cainozoic Palaeoclimates of the southern hemisphere, pp. 445-454. Rotterdam, A. A. Balkema.

Besler, H., and Marker, M. E. 1979. Namib Sandstone: A distinct lithological unit. Transactions of the Geological Society of South Africa 82:155-160.

Beukes, N. J. 1970. Stratigraphy and sedimentology of the Cave Sandstone Stage, Karroo System. In S. H. Haughton (ed.), Proceedings of the Second International Union of Geological Sciences Symposium on Gondwana Stratigraphy and Palaeontology, pp. 321-341. Pretoria, Council for Scientific and Industrial Research.

Blackwelder, E. 1928. Mudflows as a geologic agent in semi-arid mountains. Bulletin of the Geological Society of America 39:465-484.

Blakey, R. C., and Gubitosa, R. 1984. Controls of sandstone body geometry and architecture in the Chinle Formation (Upper Triassic), Colorado Plateau. Sedimentary Geology 38:51-86.

Blatt, H.; Middleton, G.; and Murray, R. 1980. Origin of sedimentary rocks. Englewood Cliffs, N.J., Prentice-Hall. 782 pp.

Blümel, W.D. 1976. Kalkkrustenvorkommen in Südwestafrika; Untersuchungsmethoden und ihrer Aussage. Mitteilungen, Basler Afrika Bibliographien 15:17-50.

_____. 1979. Zur Struktur, Reliefgebundenheit und Genese südwestafrikanischer und südostspanischer Kalkkrusten. Zeitschrift für Geomorphologie, Supplementband 33:154-167.

_____. 1982. Calcretes in Namibia and SE-Spain: Relations to substratum soil formation and geomorphic factors. In D. H. Yaalon (ed.), Aridic soils and geomorphic processes, Catena, Supplement no. 1, pp. 67-82.

Bowin, C.; Warsi, W.; and Milligan, J. 1981. Free-air gravity anomaly map of the world. Boulder, Colo., Geological Society of America.

Bowler, J.M., and Wasson, R. J. 1984. Glacial age environments of inland Australia. In J. C. Vogel (ed.), Late Cainozoic palaeoclimates of the southern hemisphere, pp. 183-208. Rotterdam, A. A. Balkema.

Boyle, D. R. 1984. The genesis of surficial uranium deposits. In International Atomic Energy Agency, Surficial Uranium Deposits. Technical Document, International Atomic Energy Agency IAEA-TECDOC-322: 45-52.

Brain, C. K. 1985. Interpreting early hominid death assemblages: The rise of taphonomy since 1925. In P. V. Tobias (ed.), Hominid evolution: Past, present and future, pp. 41-46. New York, Alan R. Liss.

Brain, C. K., and Brain, V. 1977. Microfaunal remains from Mirabib: Some evidence of palaeoecological changes in the Namib. Madoqua 10:285-293.

Breed, C. S.; Fryberger, S. C.; Andrews, S.; McCauley, C.; Lennartz, F.; Gebel, D.; and Horstman, K. 1979. Regional studies of sand seas using Landsat (ERTS) imagery. In E. D. McKee (ed.), A study of global sand seas. Professional Paper, United States Geological Survey, no. 1052, pp. 305-397.

Bridge, J. S., and Leeder, M. R. 1979. A simulation model of alluvial stratigraphy. Sedimentology 26:617-644.

Briot, P. 1984. Surficial uranium deposits in Somalia. In International Atomic Energy Agency, Surficial Uranium Deposits. Technical Document, International Atomic Energy Agency IAEA-TECDOC-322: 217-220.

Brundrit, G. B. 1981. Upwelling fronts in the southern Benguela region. Transactions of the Royal Society of South Africa 44:309-313.

Bryan, K. 1925. The Papago country, Arizona. Water Supply Paper, U. S. Geological Survey, no. 499. 436 pp.

Büdel, J. 1957. Die "Doppelten Einebnungsflächen" in den feuchten Tropen. Zeitschrift für Geomorphologie, N. F. 1:201-228.

_____. 1977. Climatic geomorphology. Princeton, Princeton University Press. 443 pp.

Bull, W. B. 1972. Recognition of alluvial-fan deposits in the stratigraphic record. In R. W. K. Hamblin and J. K. Rigby (eds.), Recognition of ancient sedimentary environments, Special Publication, Society of Economic Paleontologists and Mineralogists, no. 16, pp. 63-83.

_____. 1977. The alluvial fan environment. Progress in Physical Geography 1:222-270.

_____. 1979. Threshold of critical power in streams. Bulletin of the Geological Society of America 90(1):453-464.

_____. 1988. Floods: degradation and aggradation. In V. R. Baker, R. C. Kochel, and P. C. Patton (eds.), Flood Geomorphology, ch. 10. New York, Wiley Interscience.

Busche, D., and Hagedorn, H. 1980. Landform development in warm deserts. Zeitschrift für Geomorphologie, Supplementband 36:123-139.

Butzer, K. W. 1973. Pluralism in geomorphology. Proceedings of the Association of American Geographers 5:439-443.

_____. 1974. Geological and ecological perspectives on the Middle Pleistocene. Quaternary Research 4:136-148.

_____. 1976a. Lithostratigraphy of the Swartkrans Formation. South African Journal of Science 72:136-141.

_____. 1976b. Geomorphology from the earth. New York, Harper and Row. 463 pp.

_____. 1976c. Pleistocene climates. Geoscience and Man 13:27-44.

_____. 1978. Climate patterns in an unglaciated continent. Geographical Magazine 51:201-208.

_____. 1984. Late Quaternary environments in South Africa. In J. C. Vogel (ed.), Late Cainozoic palaeoclimates of the southern hemisphere, pp. 235-264. Rotterdam, A. A. Balkema.

Butzer, K. W.; Stuckenrath, R.; Bruzewicz, A. J.; and Helgren, D. M. 1978. Late Cenozoic paleoclimates of the Gaap Escarpment, Kalahari margin, South Africa. Quaternary Research 10:310-339.

Cagle, F. R. 1975. Evaporite deposits of the Central Namib Desert, Namibia. M.S. thesis, University of New Mexico. 155 pp.

Cambell, C. V. 1976. Reservoir geometry of a fluvial sheet sandstone. Bulletin of the American Association of Petroleum Geologists 60:1009-1020.

Cant, D. J. 1978. Development of a facies model for sandy braided river sedimentation: Comparison of the South Saskatchewan River and the Battery Point Formation. In A. D. Miall (ed.), Fluvial sedimentology. Memoir, Canadian Society of Petroleum Geologists, no. 5, pp. 627-639.

Carlisle D. 1984. Surficial uranium occurrences in relation to climate and physical setting. In International Atomic Energy Agency, Surficial Uranium Deposits. Technical Document, International Atomic Energy Agency IAEA-TECDOC-322: 25-35.

Carrington, A. J., and Kensley, B. F. 1969. Pleistocene molluscs from the Namaqualand coast. Annals of the South African Museum 52:189-223.

Chorley, R. J.; Schumm, S. A.; and Sugden, D. D. 1984. Geomorphology. London, Methuen. 605 pp.

Christiansen, F. W. 1963. Polygonal fracture and fold systems in the salt crust, Great Salt Lake Desert, Utah. Science 139:607-609.

Clemmensen, L. B., and Abrahamsen, K. 1983. Aeolian stratification and facies association in desert sediments, Arran basin (Permian), Scotland. Sedimentology 30:311-339.

CLIMAP Project Members. 1976. The surface of ice-age earth. Science 191:1131-1137.

Cockcroft, M. J.; Wilkinson, M. J.; and Tyson, P. D. 1987. The application of a present-day climatic model to the Late Quaternary in southern Africa. Climatic Change 10:161-181.

Cody, R. D. 1979. Lenticular gypsum: Occurrences in nature, and experimental determinations of effects of soluble green plant material on its formation. Journal of Sedimentary Petrology 49:1015-1028.

Coetzee, J. A. 1978. Late Cainozoic paleoenvironments of southern Africa. In E. M. van Zinderen Bakker (ed.), Antarctic glacial history and world palaeoenvironments, pp. 115-127. Rotterdam, A. A. Balkema.

_____. 1980. Tertiary environmental changes along the south-west African coast. Palaeontologia Africana 23:197-203.

Coker, W. B., and Dilabio, R. N. W. 1979. Initial geochemical results and exploration significance of two uraniferous peat bogs, Kasmere Lake, Manitoba. Paper, Geological Survey of Canada, no. 79-18, pp. 199-206.

Collinson, J. D. 1978. Vertical sequence and sand body shape in alluvial sequences. In A. D. Miall (ed.), Fluvial sedimentology. Memoir, Canadian Society of Petroleum Geologists, no. 5, pp. 577-586.

Cooke, R. U., and Warren, A. 1973. Geomorphology in deserts. London, Batsford. 394 pp.

Cooper, W. S. 1958. Coastal sand dunes of Oregon and Washington. Memoir, Geological Society of America, no. 72. 169 pp.

Coque, R. 1962. La Tunisie pre-Saharienne: Etude géomorphologique. Paris, Colin. 476 pp.

Corvinus, G., and Hendey, Q. B. 1978. A new Miocene vertebrate locality at Arrisdrift in Namibia (South West Africa). Monatshefte, Neues Jahrbuch für Geologie und Paläontologie, pt. 3, pp. 193-205.

Currie, R. 1953. Upwelling in the Benguela Current. Nature 171:497-500.

Davies, O. 1973. Pleistocene shorelines in the western Cape and South West Africa. Annals of the Natal Museum 21:719-765.

Davis, W. M. 1938. Sheetfloods and streamfloods. Bulletin of the Geological Society of America 49:1337-1416.

Deacon, H. J. 1983. Another look at the Pleistocene climates of South Africa. South African Journal of Science 79:325-328.

Deacon, J., and Lancaster, N. 1984. A synthesis of the evidence for climatic changes in southern Africa over the last 125,000 years. Report, National Weather Programme for Weather, Climate and the Atmosphere, Pretoria, Council for Scientific and Industrial Research (CSIR). 283 pp.

_____. 1988. Late Quaternary palaeoenvironments of southern Africa. Oxford, Clarendon Press.

Deacon, J.; Lancaster, N.; and Scott, L. 1984. Evidence for Later Quaternary climatic change in southern Africa: Summary of the proceedings of the SASQUA Workshop, Johannesburg, September 1983. In J. C. Vogel (ed.), Late Cainozoic palaeoclimates of the southern hemisphere, pp. 391-404. Rotterdam, A. A. Balkema.

De Martonne, E. 1927. Regions of interior basin drainage. Geographical Review 17:397-414.

D'Hoore, J. L. 1965. Soil map of Africa, explanatory monograph. Joint Project, Commission for Technical Cooperation in Africa (Lagos), no. 2. 205 pp.

Diester-Haas, L., and Schrader, H.-J. 1979. Neogene coastal upwelling history off northwest and southwest Africa. Marine Geology 29:39-53.

Dingle, R. V. 1979. Sedimentary basins and basement structures on the continental margin of South Africa. Bulletin, Papers on Marine Geoscience (Pretoria) 63:29-45.

Dingle, R. V.; Siesser, W. G.; and Newton, A. R. 1983. Mesozoic and Tertiary geology of southern Africa. Rotterdam, A. A. Balkema. 375 pp.

Dott, R. H. 1983. 1982 SEPM presidential address: Episodic sedimentation—How normal is average? How rare is rare? Does it matter? Journal of Sedimentary Petrology 53:5-23.

Dresch, J. 1957. Pediments et glacis d'erosion, pediplains et inselbergs. L'information Géographique 22:183-196.

Dubief, J. 1953. Les vents de sable dans le Sahara français. Actions Eoliennes, Centre Nationale de Recherches Scientifique (Paris), Collection International 35:45-70.

Einsele, G., and Seilacher, A. 1982. Cyclic and event stratification. Berlin, Springer-Verlag. 536 pp.

Embley, R. W., and Moreley, J. J. 1980. Quaternary sedimentation and palaeoenvironmental studies off Namibia (South West Africa). Marine Geology 36:183-204.

Erhart, H. 1967. La Genèse des sols en tant que phénomène géologique. 2d ed. Paris, Masson. 177 pp.

Eriksson, E. 1958. The chemical climate and saline soils in the arid zone. UNESCO Arid Zone Research 10:147-180.

Eriksson, P. G. 1978. An investigation of Quaternary aeolian-lacustrine sediments in Namaqualand. Palaeoecology of Africa 10:41-46.

_____. 1979. Mesozoic sheetflow and playa sediments of the Clarens Formation in the Kamberg area of the Natal Drakensberg. Transactions of the Geological Society of South Africa 82:257-258.

_____. 1981. A palaeoenvironmental analysis of the Clarens Formation in the Natal Drakensberg. Transactions of the Geological Society of South Africa 84:7-18.

Estes, R. 1978. Relationships of the South African fossil frog *Euxenopoides reuningi* (Anura, Pipidae). Annals of the South African Museum 73:49-80.

Eyles, N.; Eyles, C. H.; and Miall, A. D. 1984. Lithofacies types and vertical profile models: An alternative approach to the description and environmental interpretation of glacial diamict and diamictite sequences. Sedimentology 30:191-209.

Flohn, H. 1984. Climate evolution in the southern hemisphere and the equatorial region during the late Cenozoic. In J. C. Vogel (ed.), Late Cainozoic palaeoclimates of the southern hemisphere, pp. 5-20. Rotterdam, A.A. Balkema.

Food and Agriculture Organization of the United Nations. 1977. FAO-UNESCO soil map of the world, vol 6, Africa. Paris, United Nations Educational, Scientific and Cultural Organization. 299 pp.

Franz, H. 1970. Die gegenwärtige Insektenverbreitung und ihre Entstehung. In H. Franz and M. Beier (eds.), Die geographische Verbreitung der Insekten, Handbuch der Zoologie, sect. 4, pt. 2, pp. 1-139.

Friend, P. F. 1978. Distinctive features of some ancient river systems. In A. C. Miall (ed.), Fluvial sedimentology. Memoir, Canadian Society of Petroleum Geologists, no. 5, pp. 531-542.

_____. 1983. Towards the field classification of alluvial architecture or sequence. In J. D. Collinson and J. Lewin (eds.), Modern and ancient fluvial systems. Special Publication, International Association of Sedimentologists, no. 6, pp. 345-354.

Friend, P. F.; Slater, M. J.; and Williams, R. C. 1979. Vertical and lateral building of river sandstone bodies, Ebro basin, Spain. Journal of the Geological Society 136:39-46.

Fryberger, S. G. 1979. Dune forms and wind regime. In E. D. McKee (ed.), A study of global sand seas. Professional Paper, United States Geological Survey, no. 1052, pp. 137-169.

Fuller, A. O. 1985. A contribution to the conceptual modelling of pre-Devonian fluvial systems. Transactions of the Geological Society of South Africa 88:189-194.

Galloway, W. E. 1981. Depositional architecture of Cenozoic gulf coastal plain alluvial systems. In F. G. Ethridge and R. M. Flores (eds.), Recent and ancient nonmarine depositional environments: Models for exploration. Special Publication, Society of Economic Paleontologists and Mineralogists, no. 31, pp. 127-155.

Galloway, W. E., and Hobday, D. K. 1983. Terrigenous clastic depositional systems. Berlin, Springer-Verlag. 423 pp.

Galton, F. 1853. The narrative of an explorer in tropical South Africa. London, John Murray.

Gardner, R., and Pye, K. 1981. Nature, orgin and palaeoenvironmental significance of red coastal and desert dune sands. Progress in Physical Geography 5:514-534.

Geological Survey. 1964. Geological map of South West Africa. Four sheets, 1:1,000,000. Pretoria and Cape Town, Government Printer.

Geological Survey of the Republic of South Africa and South West Africa/Namibia. 1980. Geological map of South West Africa/Nambia 1:1,000,000. Pretoria, Government Printer.

Gevers, T. W. 1936. The morphology of western Damaraland and the adjoining Namib Desert of South West Africa. South African Geographical Journal 19:61-79.

Gevers, T. W., and van der Westhuyzen, J. P. 1931. The occurrences of salt in the Swakopmund area, South-West Africa. Transactions of the Geological Society of South Africa 34:61-80.

Gile, L. H.; Peterson, F. F.; and Grossman, R. B. 1965. The K-horizon: A master soil horizon of carbonate accumulation. Soil Science 99:74-82.

_____. 1966. Morphological and genetic sequences of carbonate accumulation in desert soils. Soil Science 101:317-360.

Gile, L. H.; Hawley, J. W.; and Grossman, R. B. 1981. Soils and geomorphology in the Basin and Range area of southern New Mexico. Guidebook to the Desert Project. Memoir, New Mexico Bureau of Mines and Mineral Resources, no. 39. 222 pp.

Glennie, K. W. 1970. Desert sedimentary environments. Developments in Sedimentology 14 222 pp.

Goudie, A. S. 1972. Climate, weathering, crust formation, dunes, and fluvial features of the Central Namib Desert, near Gobabeb, South West Africa. Madoqua 1:15-31.

_____. 1973. Duricrusts in tropical and subtropical landscapes. London, Oxford University Press. 174 pp.

Graham, J. R. 1983. Analysis of the Upper Devonian Munster basin, an example of a fluvial distributary system. In J. D. Collinson and J. Lewin (eds.), Modern and ancient fluvial systems. Special Publication, International Association of Sedimentologists, no. 6, pp. 473-483.

Greeley, R., and Iversen J. D. 1985. Wind as a geological process. London, Cambridge University Press. 333 pp.

Guthrie, R. L., and Witty, J. E. 1982. New designations for soil horizons and layers and the new Soil Survey Manual. Journal of the Soil Science Society of America 46:443-444.

Hallam, C. D. 1964. The geology of the coastal diamond deposits of South Africa. In S. H. Haughton (ed.), The geology of some ore deposits in southern Africa. Geological Society of South Africa 2:671-728.

Hambleton-Jones, B. B., 1976. Some fundamental concepts relating to the geology and geochemistry of uranium in desert environments. 28 pp.

_____. 1980. Preliminary report on the geology of a calcrete occurrence in South West Africa. Atomic Energy Board, Republic of South Africa, Report PER-52, 14 pp. Mimeograph.

_____. 1984. Surficial uranium deposits in Namibia. In International Atomic Energy Agency, Surficial Uranium Deposits. Technical Document, International Atomic Energy Agency, IAEA-TECDOC-322: 205-216.

Hambleton-Jones, B. B.; Levin, M.; and Wagener, G. F. 1986. Uraniferous surficial deposits in southern Africa. In C. R. Anheusser and S. Maske (eds.), Mineral deposits of southern Africa, vol. 1, pp. 2269-2287. Johannesburg, Geological Society of South Africa.

Hampton, M. A. 1975. Competence of fine-grained debris flows. Journal of Sedimentary Petrology 45:834-844.

Harms, J. C.; Southard J. B.; Spearing D. R.; and Walker R. G. 1975. Depositional environments as interpreted from primary sedimentary structures and stratification features. Short course, Society of Economic Paleontologists and Mineralogists, no. 2. 161 pp.

Harmse, H. J. von M. 1978. Schematic soil map of southern Africa south of latitude 16°30'S. In M. J. A. Werger (ed.), Biogeography and ecology of southern Africa, part 1, pp. 71-75 (Monographiae Biologicae vol. 31, pt. 1). The Hague, W. Junk.

Harmse, J. T. 1982. Geomorphologically effective winds in the northern part of the Namib sand desert. South African Geographer 10:43-52.

Harrison, M. S. J. 1984. A generalized classification of South African summer rain-bearing synoptic systems. Journal of Climatology 4:547-560.

_____. 1986. A synoptic climatology of South African rainfall variability. Ph.D. thesis, University of the Witwatersrand, Johannesburg.

_____. 1988. The components of analogue concepts of southern African Quaternary climate variations: A critique. Palaeoecology of Africa 19:283-292.

Haughton, S. H. 1932. On the phosphate deposits near Langebaanweg, Cape Province. Transactions of the Geological Society of South Africa 35:119-124.

Heine, K. 1982. The main stages of the Late Quaternary evolution of the Kalahari region, southern Africa. Palaeoecology of Africa 15:53-76.

Heine, K., and Geyh, M. A. 1984. Radiocarbon dating of speleothems from the Rössing Cave, Namib Desert, and palaeoclimatic implications. In J. C. Vogel (ed.), Late Cainozoic palaeoclimates of the southern hemisphere, pp. 465-470. Rotterdam, A. A. Balkema.

Hendey, Q. B. 1976. The Pliocene fossil occurrences in "E" quarry, Langebaanweg, South Africa. Annals of the South African Museum 69:215-247.

Hendey, Q. B. 1978. Preliminary report on the Miocene vertebrates from Arrisdrift, South West Africa. Annals of the South African Museum 76:1-41.

_____. 1981. Palaeoecology of the late Tertiary fossil occurrences in "E" Quarry, Langebaanweg, South Africa, and a reinterpretation of their geological context. Annals of the South African Museum 84:1-104.

_____. 1983. Cenozoic geology and palaeogeography of the fynbos region. In H. J. Deacon, Q. B. Hendey, and J. J. N. Lambrechts (eds.), Fynbos palaeoecology: A preliminary synthesis. Report, South African National Scientific Programmes, no. 75, pp. 35-60.

Higgins, C. G. 1956. Formation of small ventifacts. Journal of Geology 64:506-517.

Hooke, R. LeB. 1967. Processes on arid-region alluvial fans. Journal of Geology 75:438-460.

Hopwood, A. T. 1929. New and little known mammals from the Miocene of Africa. American Museum Novitates 344:1-9.

Horne, R. R. 1975. The association of alluvial fan, aeolian and fluviatile facies in the Caherbla Group (Devonian), Dingle Peninsula, Ireland. Journal of Sedimentary Petrology 45:535-540.

Horta, J. C. de O. S. 1980. Calcrete, gypcrete and soil classification in Algeria. Engineering Geology 15:15-52.

_____. 1981. Personal communication with the author.

Hövermann, J. 1978. Formen und Formung in der Pränamib (Flächen-Namib). Zeitschrift für Geomorphologie, Supplementband 30:55-73.

Hubert, J. F., and Hyde, M. G. 1982. Sheet-flow deposits of graded beds and mudstones on an alluvial sandflat-playa system: Upper Triassic Blomidon redbeds, St. Mary's Bay, Nova Scotia. Sedimentology 29:457-474.

Hubert J. F., and Mertz, K. A. 1984. Eolian sandstones in Upper Triassic Lower Jurassic red beds of the Fundy Basin, Nova Scotia. Journal of Sedimentary Petrology 54:798-810.

Hüser, K. 1976. Kalkkrusten im Namib-Randbereich des mittleren Südwestafrika. Mitteilungen, Basler Afrika Bibliographien 15:51-77.

Innes, J. L. 1983. Debris flows. Progress in Physical Geography 7:469-501.

International Subcommission on Stratigraphic Classification (ISSC). 1976. International Stratigraphic Guide. New York, John Wiley and Sons. 200 pp.

Iriondo, M. H. 1984. The Quaternary of northeastern Argentina. Quaternary of South America 2:51-78.

Iversen, J. D. 1979. Drifting snow similitude. Journal of the Hydraulics Division of the American Society of Civil Engineering 105:737-753.

Jackson, S. P., and Tyson, P. D. 1971. Aspects of weather and climate over southern Africa. Occasional Paper, University of the Witwatersrand, Johannesburg, Department of Geography and Environmental Studies, no. 6. 13 pp.

Jaeger, F. 1965. Geographische Landschaften Südwestafrikas. Windhoek, South West Africa Scientific Society. 251 pp.

Jenny, H. 1929. Relation of temperature to the amount of nitrogen in soils. Soil Science 27:169-188.

Kaiser, E. 1926. Die Diamantenwüste Südwestafrikas. 2 vols. Dietrich Reimer, Berlin.

Karcz, I. 1973. Sedimentary structures formed by flash floods in southern Israel. Sedimentary Geology 7:161-182.

Kent, L. E. 1947. Diatomaceous deposits in the Union of South Africa with special reference to kieselguhr, part 1. Geology and economic aspects. Memoir, Geological Survey of South Africa, no. 42. 184 pp.

Kent, L. E., and Gribnitz, K.-H. 1985. Freshwater shell deposits in the northwestern Cape Province: Further evidence for a widespread wet phase during the late Pleistocene in southern Africa. South African Journal of Science 81:361-370.

King, L. C. 1967. Morphology of the earth. 2d ed. London, Oliver and Boyd. 726 pp.

Knighton, D. 1984. Fluvial forms and processes. Baltimore, Edward Arnold. 218 pp.

Knox, J. C. 1976. Concept of the graded stream. In R. C. Flemal and W. N. Melhorn (eds.), Theories of landform development. Proceedings of the Sixth Annual Geomorphology Symposium Series, pp. 169-198. Binghamton, N.Y.

Koch, C. 1961. Some aspects of abundant life in the vegetationless sand of the Namib Desert dunes. Journal of the South West Africa Scientific Society 15:8-34, 77-92.

_____. 1962. The tenebrionidae of southern Africa XXXI. Comprehensive notes on the tenebrionid fauna of the Namib Desert. Annals of the Transvaal Museum 24:61-106.

Korn, H., and Martin, H. 1957. The Pleistocene in South West Africa. In J. Desmond Clark (ed.), Proceedings of the Third Pan African Conference on Prehistory, Livingstone, Northern Rhodesia, 1955, pp. 14-22. London, Chatto and Windus.

Lancaster, J.; Lancaster, N.; and Seely, M. K. 1984. Climate of the Central Namib desert. Madoqua 14:5-61.

Lancaster, N. 1979. Quaternary environments in the arid zone of southern Africa. Occasional Paper, University of the Witwatersrand, Johannesburg, Department of Geography and Environmental Studies, no. 22. 77 pp.

_____. 1980. The formation of seif dunes from barchans—supporting evidence for Bagnold's model from the Namib Desert. Zeitschrift für Geomorphologie, N. F. 24:160-167.

_____. 1981a. Aspects of the morphometry of linear dunes of the Namib Desert. South African Journal of Science 77:366-368.

_____. 1981b. Palaeoenvironmental implications of fixed dune systems in southern Africa. Palaeogeography, Palaeoclimatology, Palaeoecology 33:327-346.

_____. 1982a. Linear dunes. Progress in Physical Geography 6:475-504.

_____. 1982b. Dunes on the Skeleton Coast, Namibia (South West Africa): Geomorphology and grain size relationships. Earth Surface Processes and Landforms 7:575-587.

_____. 1983. Controls of dune morphology in the Namib sand sea. In M. E. Brookfield and T. S. Ahlbrandt (eds.), Eolian sediments and processes, pp. 261-289. Amsterdam, Elsevier.

_____. 1984a. Aeolian sediments, processes and landforms. Journal of Arid Environments 7:31-42.

_____. 1984b. Paleoenvironments in the Tsondab valley, Central Namib Desert. Palaeoecology of Africa 16:411-419.

_____. 1985. Winds and sand movements in the Namib sand sea. Earth Surface Processes and Landforms 10:607-619.

Lancaster, N., and Ollier, C. D. 1983. Sources of sand for the Namib sand sea. Zeitschrift für Geomorphologie, Supplementband 45:71-83.

Lane, E. W. 1955. The importance of fluvial morphology in hydraulic engineering. Proceedings of the American Society of Civil Engineers 18:1-17.

Lehr, J. R.; Frazier, A. W.; and Smith, J. P. 1966. Precipitated impurities in wet-process phosphoric acid. Journal of Agricultural Food Chemistry 14:27-33.

Leopold, L. B., and Maddock, T. 1953. The hydraulic geometry of stream channels and some physiographic implications. Professional Paper, United States Geological Survey, no. 352. 57 pp.

Levinson, A. 1985. Letter dated January 12.

Lindesay, J. A.; Harrison, M. S. J.; and Haffner, M. P. 1986. The southern oscillation and South African rainfall. South African Journal of Science 82:196-198.

Lindesay, J. A., and Tyson, P. D. 1990. Thermo-topographically induced boundary layer oscillations over the Central Namib, southern Africa. International Journal of Climatology 10:63-77.

Logan, R. F. 1960. The Central Namib Desert, South West Africa. National Academy of Sciences/National Research Council, Washington, D. C., Publication 758 (ONR Field Research Program, Report 9). 162 pp.

_____. 1969. Geography of the Central Namib Desert. In W. G. McGinnies and B. Goldman (eds.), Arid lands in perspective, pp. 127-143. Tucson, University of Arizona Press.

_____. 1972. The geographical division of the deserts of South West Africa. Mitteilungen, Basler Afrika Bibliographien 4-6:46-65.

_____. 1973. The utilization of the Namib Desert, South West Africa. In D. H. K. Amiran and A. W. Wilson (eds.), Coastal deserts: Their natural and human environments, pp. 177-186. Tucson, University of Arizona Press.

Lydolph, P. E. 1973. On the causes of aridity along a selected group of coasts. In D. H. K . Amiran and A. W. Wilson (eds.), Coastal deserts: Their natural and human environments, pp. 67-72. Tucson, University of Arizona Press.

Mabbutt, J. A. 1952. The evolution of the Middle Ugab Valley, Damaraland, South West Africa. Transactions of the Royal Society of South Africa 33:333-365.

_____. 1977. Desert landforms. Canberra, Australian National University Press. 340 pp.

McGee, W J. 1897. Sheetflood erosion. Bulletin of the Geological Society of America 8:87-112.

Machel, H. G. 1985. Fibrous gypsum and fibrous anhydrite in veins. Sedimentology 32:443-454.

Machette, M. N. 1985. Calcic soils of the southwestern United States. In D. L. Weide (ed.), Soils and Quaternary geology of the Southwestern United States. Special Paper, United States Geological Survey, no. 203, pp. 1-21.

McKee, E. D. 1982. Sedimentary structures in dunes of the Namib Desert, South West Africa. Special Paper, Geological Society of America, no. 188. 64 pp.

McKee, E. D.; Crosby, E. J.; and Berryhill, H. L. 1967. Flood deposits, Bijou Creek, Colorado, June 1965. Journal of Sedimentary Petrology 37:829-851.

McKee, E. D., and Tibbitts, G. C. 1964. Primary structures of a seif dune and associated deposits in Libya. Journal of Sedimentary Petrology 34:5-17.

McPherson, J. G.; Shanmugam, G.; and Moiola R. J. 1987. Fan-deltas and braid deltas: Varieties of coarse-grained deltas. Bulletin of the Geological Society of America 99: 331-340.

_____. 1988. Fan deltas and braid deltas: Conceptual problems. In W. Nemec and R. J. Steel (eds.), Fan deltas: Sedimentology and tectonic settings, pp. 14-22. Glasgow and London, Blackie and Son.

MacVicar, C. N.; De Villiers, J. M.; Loxton, R. F.; Verster, E.; Lambrechts, J. J. N.; Merryweather, F. R.; Le Roux, J.; Van Rooyen, T. H.; and Harmse, H. J. 1977. Soil classification: A binomial system for South Africa. Pretoria, Department of Agricultural Technical Services. 150 pp.

Makkavayev, N. I. 1972. The impact of large water engineering projects on geomorphic processes in stream valleys. Soviet Geography 13:387-293.

Mallick, D. I. J.; Habgood, F.; and Skinner, A. C. 1981. A geological interpretation of LANDSAT imagery and air photography of Botswana. Overseas Geology and Mineral Resources 56:1-35.

Mann, A. W., and Deutscher, R. L. 1978. Genesis principles for the precipitation of carnotite in calcrete drainages in western Australia. Economic Geology 73:1724-1737.

Marker, M. E. 1973. Tufa formation in the Transvaal. Zeitschrift für Geomorphologie, N. F. 17:460-473.

_____. 1977. Aspects of the geomorphology of the Kuiseb River, South West Africa. Madoqua 10:199-206.

_____. 1979. Relict fluvial terraces on the Tsondab Flats, Namibia. Journal of Arid Environments 2:133-177.

_____. 1982. Aspects of Namib geomorphology: A doline karst. Palaeoecology of Africa 15:187-199.

Marker, M. E., and Besler, H. 1979. Namib sandstone: A distinct lithological unit. Transactions of the Geological Society of South Africa 82:155-160.

Marker, M. E., and Müller, D. 1978. Relict vlei silt of the middle Kuiseb Valley, South West Africa. Madoqua 2:151-162.

Martin, H. 1950. Südwest-Afrika. Geologische Rundschau 38:6-14.

_____. 1961. Abriss der geologischen Geschichte Südwestafrikas. Journal of the South West Africa Scientific Society 15:57-66.

_____. 1963. A suggested theory for the origin and a brief description of some gypsum deposits of South West Africa. Transactions of the Geological Society of South Africa 66:345-352.

_____. 1968. Paläomorphologische Formelemente in den Landschaften Südwest-Afrikas. Geologische Rundschau 58:121-128.

_____. 1974. The sheltering desert. Windhoek, South West Africa Scientific Society. 234 pp.

_____. 1975. Structural and palaeogeographical evidence for an Upper Palaeozoic sea between southern Africa and South America. Proceedings Papers, Third Gondwana Symposium, International Union of Geological Sciences, Canberra, pp. 37-51.

Martin, H., and Mason, R. 1954. The test trench in the Phillips Cave, Ameib, Erongo Mountains, South West Africa. South African Archaeological Bulletin 9:148-151.

Maull, D. J., and East, L. F. 1963. Three-dimensional flow in cavities. Journal of Fluid Mechanics 16:620-632.

Meigs, P. 1966. Geography of coastal deserts. UNESCO Arid Zone Research 28. 140 pp.

Miall, A. D., 1977. A review of the braided river depositional environment. Earth-Science Reviews 13:1-62.

_____. 1978. Lithofacies types and vertical profile models in braided river deposits: A summary. In A. D. Miall (ed.), Fluvial sedimentology. Memoir, Canadian Society of Petroleum Geologists, no. 5, pp. 597-604.

_____. 1980. Cyclicity and the facies model concept in fluvial deposits. Bulletin of the Canadian Petroleum Geology 28:59-80.

_____. 1981. Alluvial sedimentary basins: Tectonic setting and basin structure. In A. D. Miall (ed.), Sedimentation and tectonics in alluvial basins. Special Paper, Geological Association of Canada 23:1-33.

_____. 1984. Principles of sedimentary basin analysis. New York, Springer-Verlag. 490 pp.

_____. 1985. Architectural-element analysis: A new method of facies analysis applied to fluvial deposits. Earth-Science Reviews 22:261-308.

_____. 1987. Recent developments in the study of fluvial facies models. In F. G. Ethridge, R. M. Flores, and M. D. Harvey (eds.), Recent developments in fluvial sedimentology. Special Publication, Society of Economic Paleontologists and Mineralogists, no. 39, pp. 1-9.

Middleton, G. V., and Hampton, M. A. 1973. Sediment gravity flows: Mechanics of flow and deposition. In L. J. Doyle and O. H. Pilkey (eds.), Turbidites and deep-water sedimentation, Society of Economic Paleontologists and Mineralogists, Pacific Section, Short Course Lecture Notes, pp. 1-38.

Miller, R. McG., and Seely, M. K. 1976. Fluvio-marine deposits south-east of Swakopmund, South West Africa. Madoqua 9:23-26.

Moisel, A. 1975. A Braun-Blanquet survey of the vegetation of the Welwitschia Plain. University of Cape Town.

Moreley, J. J., and Hays, J. D. 1979. Comparison of glacial and interglacial oceanographic conditions in the south Atlantic from variations in calcium carbonate and radiolarian distributions. Quaternary Research 12:396-408.

Mukerji, A. B. 1976. Terminal fans of inland streams in Sutlej-Yamuna plain, India. Zeitschrift für Geomorphologie, N. F. 20:190-204.

Munsell Color Company. 1954. Munsell Soil Color Charts. Baltimore, Munsell Soil Color Company.

Nash, C. R. 1972. Primary anhydrite in precambrian gneisses from the Swakopmund District, South West Africa. Contributions to Mineralogy and Petrology 36:27-32.

Nel, P. S., and Opperman, D. J. P. 1985. Vegetation types of the Gravel Plains. In B. J. Huntley (ed.), The Kuiseb environment: The development of a monitoring baseline.

Report, South African National Scientific Programmes, Council for Scientific and Industrial Research (Pretoria), no. 106, pp. 118-125.

Neal, J. T., and Motts, W. S. 1967. Recent geomorphic changes in the playas of the western United States. Journal of Geology 75:511-525.

Needham, H. D. 1978. Color of sediments. In R. W. Fairbridge and J. Bourgeois (eds.), The encyclopedia of sedimentology, pp. 173-176. Stroudsburg, Pa., Dowden, Hutchinson and Ross.

Nemec, W., and Steel, R. J. 1988. What is a fan delta and how do we recognize it? In W. Nemec and R. J. Steel (eds.), Fan deltas: Sedimentology and tectonic settings, pp. 3-13. Glasgow and London, Blackie and Son.

Nemec, W., and Steel, R. J., eds. 1988. Fan deltas: Sedimentology and tectonic settings. Glasgow and London, Blackie and Son.

Newell, R. E.; Gould-Stewart, S.; and Chung, J. C. 1981. A possible interpretation of palaeoclimatic reconstructions for 18,000 BP for the region 60°N to 60°S, 60°W to 100°E. Palaeoecology of Africa 13:1-19.

Netterberg, F. 1969. Ages of calcrete in southern Africa. South African Archaeological Bulletin 24:88-92.

Nichols, G. 1987. Structural controls on fluvial distributary systems: The Luna system, northern Spain. In F. G. Ethridge, R. M. Flores, and M. D. Harvey (eds.), Recent developments in fluvial sedimentology. Special Publication, Society of Economic Paleontologists and Mineralogists, no. 39, pp. 269-277.

Nicholson, S. E., and Flohn, F. 1980. African environmental and climatic changes and the general atmospheric circulation in the late Pleistocene and Holocene. Climatic Change 2:313-348.

Nienaber, G. S., and Raper, P. E. 1977. Toponymica Hottentotica. Pretoria, Human Sciences Research Council, vol. 2 (H-Z). 1126 pp.

North American Commission on Stratigraphic Nomenclature. 1983. North American Stratigraphic Code. Bulletin of the American Association of Petroleum Geologists 67:841-875.

Ollier, C. D. 1977. Outline geological and geomorphic history of the Central Namib Desert. Madoqua 10:207-212.

_____. 1978. Inselbergs of the Namib Desert: Processes and history. Zeitschrift für Geomorphologie, N. F. 31:161-176.

Olsen, H. 1987. Ancient ephemeral stream deposits: A local terminal fan model from the Bunter Sandstone Formation (L. Triassic) in the Tønder-3, -4 and -5 wells, Denmark. In L. E. Frostick and I. Reid (eds.), Desert sediments: Ancient and modern. Special Publication, Geological Society (London), no. 35, pp. 69-86.

Parkash, B.; Awasthi, A. K.; and Gohain, K. 1983. Lithofacies of the Markanda terminal fan, Kurukshetra district, Haryana, India. In J. D. Collinson and J. Lewin (eds.), Modern and ancient fluvial systems. International Association of Sedimentologists, Special Publications, no. 6, pp. 337-344.

Partridge, T. C. 1985. Spring flow and tufa accretion at Taung. In P. V. Tobias (ed.), Hominid evolution: Past, present, and future, pp. 171-188. New York, Alan R. Liss.

Peterson, F. 1984. Fluvial sedimentation on a quivering craton: Influence of slight crustal movements on fluvial processes, upper Jurassic Morrison Formation, Western Colorado Plateau. Sedimentary Geology 38:21-49.

Picard, M. D., and High, L. R. 1973. Sedimentary structures of ephemeral streams. Developments in Sedimentology (Amsterdam) 17. 203 pp.

Potter, P. E. 1967. Sand bodies and sedimentary environments: A review. Bulletin of the American Association of Petroleum Geologists 51:337-365.

Proto-Decima, F.; Medizza, F.; and Todesco, L. 1978. Southeastern Atlantic Leg 40 calcareous nannofossils. Initial Reports of the Deep Sea Drilling Project 40:571-634.

Pye, K. 1983. Red beds. In A. S. Goudie, and K. Pye, (eds.), Chemical sediments and geomorphology: Precipitates and residua in the near-surface environment, pp. 227-264. London, Academic Press.

Rachocki, A. H. 1981. Alluvial fans: An attempt at an empirical approach. New York, John Wiley and Sons. 161 pp.

Rahn, P. H. 1967. Sheetfloods, streamfloods, and the formation of pediments. Annals of the Association of American Geographers 57:593-604.

Reading, H. G. (ed.) 1986. Sedimentary environments and facies. 2d ed., Oxford, Blackwell Scientific Publications. 680 pp.

Reineck, H., and Singh, I. B. 1980. Depositional sedimentary environments. 2d ed., Berlin, N.Y., Springer Verlag. 549 pp.

Reuning, E. 1925. Gediegen Schwefel in der Küstenzone Südwestafrikas. Centralblatt für Geologie and Paläontologie, Abteilung A:86-94.

Risacher, F. 1978. Genèse d'une croûte de gypse dans un basin de l'Altiplano bolivien. Cahiers d'Office de la Recherche Scientifique et Technique d'Outre Mer, Serie Geologie 10:91-100.

Robinson, E. R. 1977. A plant ecological study of the Namib Desert Park. M.S. thesis, University of Natal, Pietermaritzburg.

Royal Navy and South African Air Force. 1944. Weather on the coasts of southern Africa 2, part 1, pp. 1-61 (Meteorological Services of the Royal Navy and the South African Airforce).

Rust, B. R. 1978. Depositional models for braided alluvium. In A. D. Miall (ed.), Fluvial sedimentology. Memoir, Canadian Society of Petroleum Geologists, no. 5, pp. 605-626.

_____. 1981. Sedimentation in an arid-zone anastomosing fluvial system: Cooper's Creek, central Australia. Journal of Sedimentary Petrology 51:745-755.

Rust, B. R., and Gibling, M. R. 1990. Braidplain evolution in the Pennsylvanian South Bar Formation, Sydney Basin, Nova Scotia. Journal of Sedimentary Petrology 60: 59-72.

Rust, B. R., and Koster, E. H. 1984. Coarse alluvial deposits. In R. G. Walker (ed.), Facies models. Geoscience Canada, no. 1, pp. 53-69. Toronto, Geological Association of Canada. Reprint.

Rust, B. R., and Legun, A. S. 1983. Modern anastomosing fluvial deposits in arid central Australia, and a Carboniferous analogue in New Brunswick, Canada. In J. D. Collinson and J. Lewin (eds.), Modern and ancient fluvial systems. Special Publication, International Association of Sedimentologists, no. 6, pp. 385-392.

Rust, U. 1970. Beiträge zum Problem der Inselberglandschaften aus dem mittleren Südwestafrika. Hamburger Geographische Studien 23. 280 pp.

_____. 1975. Das Spektrum der geomorphologischen Milieus und die Relieftypendifferenzierung in der Zentralen Namib. Würzburger Geographische Arbeiten 43:79-110.

_____. 1979. Über Konvergenzen im Wüstenrelief am Beispiel der südwestafrikanischen Namibwüste (Skelettküste und Zentrale Namib). Mitteilungen, Geographische Gesellschaft München 64:201-216.

_____. 1980. Models in geomorphology: Quaternary evolution of the actual pattern of coastal central and northern Namib Desert. Palaeontologica Africana 23:173-184.

_____. 1987. Geomorphologische Forschungen im Südwestafrikanischen Kaokoveld zum angeblichen vollariden Quatären Kernraum der Namibwüste. Erdkunde 42:118-133.

_____. 1989. (Paläo)-Klima und Relief: Das Reliefgefüge der südwestafrikanischen Namibwüste (Kunene bis 27° s.B.). Münchener Geographische Abhandlungen B7. 158 pp.

Rust, U., and Schmidt, H. H. 1981. Der Fragenkreis jungkwartärer Klimaschwankungen im südwestafrikanischen Sektor des heute ariden südlichen Afrika. Mitteilungen, Geographische Gesellschaft München 66:138-174.

Rust, U.; Schmidt, H. H.; and Dietz, K. R. 1984. Paleoenvironments of the present day arid south western Africa 30,000-5000 B.P.: Results and problems. Palaeoecology of Africa 16:109-148.

Rust, U., and Vogel, J. C. 1988. Late Quaternary environmental changes in the northern Namib Desert as evidenced by fluvial landforms. Palaeoecology of Africa 19:127-137.

Rust, U., and Wieneke, F. 1974. Studies on gramadulla formation in the middle part of the Kuiseb River, South West Africa. Madoqua 3:5-15.

_____. 1976. Geomorphologie der küstennahen Zentralen Namib (Südwestafrika). Münchener Geographische Abhandlungen 19. 74 pp.

Sandelowsky, B. H. 1977. Mirabib: An archaeological study in the Namib. Madoqua 10:221-283.

Sauer, E. G. F. 1972. Fund eines Nashorn-Vorderhorns in der zentralen Namib. Namib und Meer 3:21-23.

Sawyer, E. W. 1981. Damaran structural and metamorphic geology of an area southeast of Walvis Bay, South West Africa/Namibia. Memoir, Geological Survey of South West Africa (Pretoria), no. 7. 94 pp.

Schaller, D. 1986. Personal communication.

Scholz, H. 1963. Studien über die Bodenbildung zwischen Rehoboth und Walvis-Bay. Ph.D. thesis, University of Bonn.

_____. 1968. Die Boden der Wüste Namib/Südwestafrika. Zietschrift für Pflanzenernährung, Düngung und Bodenkunde 119:91-107.

_____. 1972. The soils of the Central Namib Desert with special consideration of the soils in the vicinity of Gobabeb. Madoqua 1:33-51.

Schultz, A. W. 1984. Subaerial debris-flow deposition in the Upper Paleozoic Cutler Formation, western Colorado. Journal of Sedimentary Petrology 54:759-772.

Schulze, R. E., and McGee, O. S. 1978. Climatic indices and classifications in relation to the biogeography of southern Africa. In M. J. A. Werger (ed.), Biogeography and Ecology of southern Africa, part 1, pp. 19-52 (Monographiae Biologicae vol. 31, pt. 1). The Hague, W. Junk.

Schumm, S. A. 1976. Episodic erosion: A modification of the geomorphic cycle. In R. Flemal and W. Melhorn (eds.), Theories of landform development: Publications on geomorphology, pp. 69-85. Binghamton, N.Y., SUNY.

_____. 1977. The fluvial system. New York, John Wiley and Sons. 338 pp.

_____. 1979. Geomorphic thresholds: The concept and its applications. Transactions of the Institute of British Geographers 4:485-515.

_____. 1981. Evolution and response of the fluvial system: Sedimentologic implications. In E. G. Ethridge and R. M. Flores (eds.), Special Publication, Society of Economic Paleontologists and Mineralogists, no. 31, pp. 19-29.

Schumm, S. A., and Hadley, R. F. 1957. Arroyos and the semiarid cycle of erosion. American Journal of Science 255:161-174.

Schumm, S. A., and Lichty, R. W. 1965. Time, space and causality in geomorphology. American Journal of Science 263:110-119.

Scott, L., and Vogel, J. C. 1983. Late Quaternary pollen profile from the Transvaal Highveld, South Africa. South African Journal of Science 79:266-272.

Seely, M. K. 1987. Personal communication with author.

Seely, M. K., and Hamilton, W. J. 1976. Fog catchment sand trenches constructed by tenebrionid beetles, Lepidochora, from the Namib Desert. Science 193:484-486.

Seely, M. K., and Sandelowsky, B. H. 1974. Dating the regression of a river's end point. South African Archaeological Society, Goodwin Series 2:61-64.

Selby, M. J. 1977a. Bornhardts of the Namib desert. Zeitschrift für Geomorphologie, N. F.. 21:1-13.

_____. 1977b. Paleowind directions in the Central Namib Desert, as indicated by ventifacts. Madoqua 10:195-198.

Selby, M. J.; Hendey, C. H.; and Seely, M. K. 1979. A late Quaternary lake in the Central Namib Desert, southern Africa, and some implications. Paleogeography, Paleoclimatology, and Paleoecology 26:37-41.

Shackleton, J. N., and Kennett, J. P. 1975. Paleotemperature history of the Cenozoic and the initiation of Antarctic glaciation: Oxygen and carbon isotope analyses in DSDP sites 277, 279 and 281. Initial Reports of the Deep Sea Drilling Project 29:743-755.

Shackley, M. 1980. An Acheulian industry with *Elephas* fauna from Namib IV, South West Africa (Namibia). Nature 284:340-341.

Siesser, W. G. 1972. Petrology of some South African coastal limestones. Transactions of the Geological Society of South Africa 75:177-185.

_____. 1978. Aridification of the Namib Desert: Evidence from oceanic cores. In E. M. Van Zinderen Bakker (ed.), Antarctic glacial history and world palaeoenvironments, pp. 105-113. Rotterdam, A. A. Balkema.

Siesser, W. G., and Dingle, R. V. 1981. Tertiary sea level movements around southern Africa. Journal of Geology 89:523-536.

Siesser, W. G., and Salmon, D. 1979. Eocene maritime sediments in the Sperrgebiet. Annals of the South African Museum 79:9-34.

Smith, D. A. M. 1965. The geology of the area around the Khan and Swakop Rivers in South West Africa. Memoir, Geological Survey of South Africa, South West Africa Series, no. 3, 113 pp.

Smoot, J. P. 1983. Depositional subenvironments in an arid closed basin; the Wilkins Peak Member of the Green River Formation (Eocene), Wyoming, U.S.A. Sedimentology 30:801-827.

Sneh, A. 1983. Desert stream sequence in the Sinai Peninsula. Journal of Sedimentary Petrology 53:1271-1280.

Soil Survey Staff. 1975. Soil taxonomy. Agriculture handbook, Soil Conservation Service, United States Department of Agriculture, no. 436. 754 pp.

South African Committee for Stratigraphy (SACS). 1980. Stratigraphy of South Africa, part 1: Lithostratigraphy of the Republic of South Africa, South West Africa/Namibia and the Republics of Bophutatswana, Transkei and Venda. Handbook, Geological Survey of South Africa, no. 8. 622 pp.

Spreitzer, H. 1965-66. Beobachtungen zur Geomorphologie der zentralen Namib und ihrer Randgebiete. Journal of the South West Africa Scientific Society 20:69-94.

Stengel, H. W. 1964. Die Riviere der Namib und ihr Zulauf zum Atlantik. Teil I: Kuiseb und Swakop. Scientific Papers, Namib Desert Research Station, no. 22. 50 pp.

_____. 1970. Die riviere van die Namib met hulle toelope na die Atlantiese Oseaan. Derde deel: Tsondab, Tsams en Tsauchab. Report, Department of Water Affairs, Windhoek, Namibia.

Stocken, C. G. 1975. Report on a visit to the uranium prospect. Internal Report, Anglo American Corporation of South Africa, Johannesburg.

Stocken, C. G., and Campbell, D. 1982. Some notes on the CDM raised beach complex. Winter Field School, Geological Society of South Africa, no. 8.

Stokes, W. L. 1950. Pediment concept applied to Shinarump and similar conglomerates. Bulletin, Geological Society of America 61:91-98.

Sweeting, M. M., and Lancaster, N. 1982. Solutional and wind erosion forms on limestone in the Central Namib Desert. Zeitschrift für Geomorphologie, N. F. 26:197-207.

Tankard, A. J. 1974a. Petrology and origin of the phosphorite and aluminium phosphate rock of the Langebaanweg-Saldanha area, south-western Cape Province. Annals of the South African Museum 65:217-249.

_____. 1974b. Varswater Formation of the Langebaanweg-Saldanha area, Cape Province. Transactions of the Geological Society of South Africa 77:265-283.

Tankard, A. J., and Rogers, J. 1978. Late Cenozoic palaeoenvironments on the west coast of southern Africa. Journal of Biogeography 5:319-337.

Tankard, A. J.; Jackson, M. P.; Eriksson, K. A.; Hobday, D. K.; Hunter, D. R.; and Minter, W. E. L. 1982. Crustal evolution of South Africa: 3.8 billion years of earth history. Berlin, Springer Verlag. 523 pp.

Thomas, P., and Veverka, J. 1979. Seasonal and secular variation of wind streaks on Mars: An analysis of Mariner 9 and Viking data. Journal of Geophysical Research 84:8131-8146.

Toens, P. D., and Hambleton-Jones, B. B. 1984. Definition and classification of superficial uranium deposits. In International Atomic Energy Agency, Surficial Uranium Deposits. Technical Document, International Atomic Energy Agency IAEA-TECDOC-322:9-13.

Tsoar, H. 1978. The dynamics of longitudinal dunes. Final Technical Report, European Research Office, United States Army. 171 pp.

Tucker, M. E. 1978. Gypsum crusts (gypcrete) and patterned ground from northern Iraq. Zeitschrift für Geomorphologie, N. F. 22.89-100.

Tunbridge, I. P. 1981. Sandy high-energy flood sedimentation: Some criteria for recognition, with an example from the Devonian of S.W. England. Sedimentary Geology 28:78-96.

_____. 1984. Facies model for a sandy ephemeral stream and clay playa complex: The Middle Devonian Trentishoe Formation of north Devon, U.K. Sedimentology 31:697-716.

Tyson, P. D. 1986. Climatic change and variability in southern Africa. Cape Town, Oxford University Press. 220 pp.

Tyson, P. D., and Seely, M. K. 1980. Local winds over the Central Namib desert. South African Geographical Journal 62:136-150.

VanArsdale, R. 1982. Influence of calcrete on the geometry of arroyos near Buckeye, Arizona. Bulletin of the Geological Society of America 93:20-26.

Van Dijk, D. E. 1978. Trackways in the Stormberg. Palaeontologica Africana 21:113-120.

Van Wyck, F. 1969. Gypsum deposits. Report, Geological Survey of South West Africa, Windhoek.

Van Zinderen Bakker, E. M. 1967. Upper Pleistocene and Holocene stratigraphy and ecology on the basis of vegetation changes in sub-Saharan Africa. In W. W. Bishop and J. D.

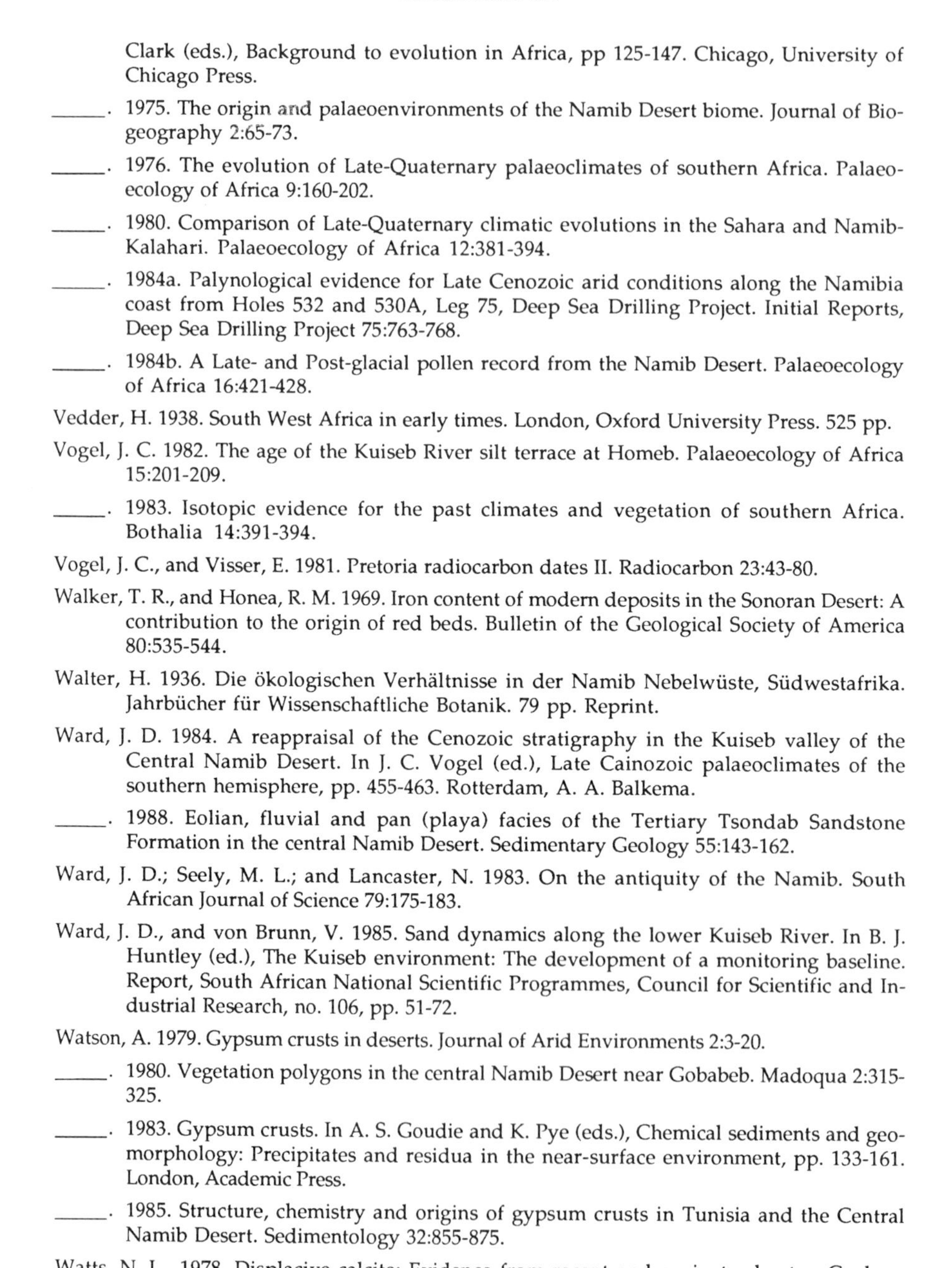

Clark (eds.), Background to evolution in Africa, pp 125-147. Chicago, University of Chicago Press.

_____. 1975. The origin and palaeoenvironments of the Namib Desert biome. Journal of Biogeography 2:65-73.

_____. 1976. The evolution of Late-Quaternary palaeoclimates of southern Africa. Palaeoecology of Africa 9:160-202.

_____. 1980. Comparison of Late-Quaternary climatic evolutions in the Sahara and Namib-Kalahari. Palaeoecology of Africa 12:381-394.

_____. 1984a. Palynological evidence for Late Cenozoic arid conditions along the Namibia coast from Holes 532 and 530A, Leg 75, Deep Sea Drilling Project. Initial Reports, Deep Sea Drilling Project 75:763-768.

_____. 1984b. A Late- and Post-glacial pollen record from the Namib Desert. Palaeoecology of Africa 16:421-428.

Vedder, H. 1938. South West Africa in early times. London, Oxford University Press. 525 pp.

Vogel, J. C. 1982. The age of the Kuiseb River silt terrace at Homeb. Palaeoecology of Africa 15:201-209.

_____. 1983. Isotopic evidence for the past climates and vegetation of southern Africa. Bothalia 14:391-394.

Vogel, J. C., and Visser, E. 1981. Pretoria radiocarbon dates II. Radiocarbon 23:43-80.

Walker, T. R., and Honea, R. M. 1969. Iron content of modern deposits in the Sonoran Desert: A contribution to the origin of red beds. Bulletin of the Geological Society of America 80:535-544.

Walter, H. 1936. Die ökologischen Verhältnisse in der Namib Nebelwüste, Südwestafrika. Jahrbücher für Wissenschaftliche Botanik. 79 pp. Reprint.

Ward, J. D. 1984. A reappraisal of the Cenozoic stratigraphy in the Kuiseb valley of the Central Namib Desert. In J. C. Vogel (ed.), Late Cainozoic palaeoclimates of the southern hemisphere, pp. 455-463. Rotterdam, A. A. Balkema.

_____. 1988. Eolian, fluvial and pan (playa) facies of the Tertiary Tsondab Sandstone Formation in the central Namib Desert. Sedimentary Geology 55:143-162.

Ward, J. D.; Seely, M. L.; and Lancaster, N. 1983. On the antiquity of the Namib. South African Journal of Science 79:175-183.

Ward, J. D., and von Brunn, V. 1985. Sand dynamics along the lower Kuiseb River. In B. J. Huntley (ed.), The Kuiseb environment: The development of a monitoring baseline. Report, South African National Scientific Programmes, Council for Scientific and Industrial Research, no. 106, pp. 51-72.

Watson, A. 1979. Gypsum crusts in deserts. Journal of Arid Environments 2:3-20.

_____. 1980. Vegetation polygons in the central Namib Desert near Gobabeb. Madoqua 2:315-325.

_____. 1983. Gypsum crusts. In A. S. Goudie and K. Pye (eds.), Chemical sediments and geomorphology: Precipitates and residua in the near-surface environment, pp. 133-161. London, Academic Press.

_____. 1985. Structure, chemistry and origins of gypsum crusts in Tunisia and the Central Namib Desert. Sedimentology 32:855-875.

Watts, N. L., 1978. Displacive calcite: Evidence from recent and ancient calcretes. Geology 6:699-703.

Wellington, J. H. 1955. Southern Africa: A geographical study, vol. 1, Physical geography. London, Cambridge University Press. 528 pp.

_____. 1967. South West Africa and its human issues. London, Oxford University Press. 461 pp.

Wells, N. A., and Dorr J. A. 1987a. Shifting of the Kosi river, northern India. Geology 15:204-207.

_____. 1987b. A reconnaissance of sedimentation on the Kosi alluvial fan of India. In F. G. Ethridge, R. M. Flores, and M. D. Harvey (eds.), Special Publication, Society of Economic Paleontologists and Mineralogists, no. 39, pp. 51-61.

Werger, M. J. A. 1978. The Karoo-Namib Region. In M. J. A. Werger (ed.), Biogeography and ecology of southern Africa, part 1, pp. 231-299 (Monographiae Biologicae, vol. 31, pt. 1). The Hague, W. Junk.

Whitehead, V. S.; Helfert, M. R.; Lulla, K. P.; Amsbury, D. L.; Wilkinson, M. J.; Stevenson, R. E.; Daley, W. J.; Johnson, W. R.; Runco, S. K.; and Muehlberger, W. R. 1990. Earth observations during space shuttle mission STS-28. Geocarto International 5(2):63-80.

Whitehouse, F. W. 1944. The natural drainage of some very flat monsoonal lands. Australian Geographer 4(7):183-196.

Wieneke, F. 1975. Entwicklung und Differenzierung des Reliefs der Küste der zentralen Namib. Würzburger Geographische Arbeiten 43:111-134.

Wieneke, F., and Rust, U. 1972. Das Satellitenbild als Hilfsmitel zur Formulierung geomorphologischer Arbeitshypothesen. Windhoek, South West Africa Scientific Society 16 pp.

_____. 1975. Zur relativen und absoluten Geochronologie der Reliefentwicklung an der Küste des mittleren Südwestafrika. Eiszeitalter und Gegenwart 26:241-250.

_____. 1976. Methodischer Ansatz, Techniken und Ergebnisse geomorphologischer Untersuchungen in der Zentralen Namib (Südwestafrika). Mitteilungen, Basler Afrika Bibliographien 15:107-150.

Wilkinson, M. J. 1976. Preliminary report on aspects of the geomorphology of the lower Tumas Basin, South West Africa. Internal Report, Anglo American Corporation of South Africa Limited, Johannesburg. 36 pp.

_____. 1977. Geomorphology and secondary uranium in the Swakopmund grant areas. Technical Report, Anglo American Corporation of South Africa Limited, Johannesburg. 26 pp.

_____. 1979. Note on three karst cavities in the Swakopmund grant areas. Internal Report, Anglo American Corporation of South Africa Limited, Johannesburg. 5 pp.

_____. 1980. Carnotite in the Swakopmund grant areas: Provenance and concentration zones. Internal Report, Anglo American Corporation of South Africa Limited, Johannesburg. 33 pp.

_____. 1981. Terrestrial origin of the gypsums of the Central Namib: An alternative hypothesis. Paper, International Soil Science Society Meeting, Jerusalem, May, 1981.

_____. 1991. "Immense alluvial cones" in Africa. Program and Abstracts, Association of American Geographers, Annual Meeting, Miami (in press).

Wilkinson, M. J.; Blaha, J. E.; and Noli, D. 1989. A new lagoon on the Namibian coast of South Africa: Sand spit growth documented from STS-29 shuttle photography. Geocarto International 1(4):63-66.

Williams, G. E. 1970. The central Australian stream floods of February-March 1967. Journal of Hydrology 11:185-200.

_____. 1971. Flood deposits of the sand-bed ephemeral streams of central Australia. Sedimentology 17:1-40.

Wilson, I. G. 1972. Aeolian bedforms: Their development and origins. Sedimentology 19:173-210.

Wopfner, H., and Twidale, C. R. 1967. Geomorphological history of the Lake Eyre basin. In J. N. Jennings and J. A. Mabbutt (eds.), Landform studies from Australia and New Guinea. pp. 118-143. Cambridge, Cambridge University Press.

Yaalon, D. H., and Ward, J. D. 1982. Observation on calcrete and recent calcic horizons in relation to landforms, Central Namib Desert. Palaeoecology of Africa 15:183-186.

Zen, E. 1965. Solubility measurements in the system $CaSO_4$-$NaCl$-H_2O at 35°, 50° and 70°C and one atmosphere pressure. Journal of Petrology 6:124-164.

Index

THE UNIVERSITY OF CHICAGO

GEOGRAPHY RESEARCH PAPERS

(Lithographed, 6 x 9 inches)

Titles in Print

127. GOHEEN, PETER G. *Victorian Toronto, 1850 to 1900: Pattern and Process of Growth.* 1970. xiii + 278 p.
130. GLADFELTER, BRUCE G. *Meseta and Campina Landforms in Central Spain: A Geomorphology of the Alto Henares Basin.* 1971. xii + 204 p.
131. NEILS, ELAINE M. *Reservation to City: Indian Migration and Federal Relocation.* 1971. x + 198 p.
132. MOLINE, NORMAN T. *Mobility and the Small Town, 1900-1930.* 1971. ix + 169 p.
133. SCHWIND, PAUL J. *Migration and Regional Development in the United States, 1950-1960.* 1971. x + 170 p.
134. PYLE, GERALD F. *Heart Disease, Cancer and Stroke in Chicago: A Geographical Analysis with Facilities, Plans for 1980.* 1971. ix + 292 p.
136. BUTZER, KARL W. *Recent History of an Ethiopian Delta: The Omo River and the Level of Lake Rudolf.* 1971. xvi + 184 p.
139. McMANIS, DOUGLAS R. *European Impressions of the New England Coast, 1497-1620.* 1972. viii + 147 p.
140. COHEN, YEHOSHUA S. *Diffusion of an Innovation in an Urban System: The Spread of Planned Regional Shopping Centers in the United States, 1949-1968.* 1972. ix + 136 p.
141. MITCHELL, NORA. *The Indian Hill-Station: Kodaikanal.* 1972. xii + 199 p.
142. PLATT, RUTHERFORD H. *The Open Space Decision Process: Spatial Allocation of Costs and Benefits.* 1972. xi + 189 p.
143. GOLANT, STEPHEN M. *The Residential Location and Spatial Behavior of the Elderly: A Canadian Example.* 1972. xv + 226 p.
144. PANNELL, CLIFTON W. *T'ai-Chung, T'ai-wan: Structure and Function.* 1973. xii + 200 p.
145. LANKFORD, PHILIP M. *Regional Incomes in the United States, 1929-1967: Level, Distribution, Stability, and Growth.* 1972. x + 137 p.
146. FREEMAN, DONALD B. *International Trade, Migration, and Capital Flows: A Quantitative Analysis of Spatial Economic Interaction.* 1973. xiv + 201 p.
147. MYERS, SARAH K. *Language Shift among Migrants to Lima, Peru.* 1973. xiii + 203 p.
148. JOHNSON, DOUGLAS L. *Jabal al-Akhdar, Cyrenaica: An Historical Geography of Settlement and Livelihood.* 1973. xii + 240 p.
149. YEUNG, YUE-MAN. *National Development Policy and Urban Transformation in Singapore: A Study of Public Housing and the Marketing System.* 1973. x + 204 p.
150. HALL, FRED L. *Location Criteria for High Schools: Student Transportation and Racial Integration.* 1973. xii + 156 p.
151. ROSENBERG, TERRY J. *Residence, Employment, and Mobility of Puerto Ricans in New York City.* 1974. xi + 230 p.
152. MIKESELL, MARVIN W., ed. *Geographers Abroad: Essays on the Problems and Prospects of Research in Foreign Areas.* 1973. ix + 296 p.
153. OSBORN, JAMES. *Area, Development Policy, and the Middle City in Malaysia.* 1974. x + 291 p.

154. WACHT, WALTER F. *The Domestic Air Transportation Network of the United States.* 1974. ix + 98 p.
155. BERRY, BRIAN J. L. et al. *Land Use, Urban Form and Environmental Quality.* 1974. xxiii + 440 p.
156. MITCHELL, JAMES K. *Community Response to Coastal Erosion: Individual and Collective Adjustments to Hazard on the Atlantic Shore.* 1974. xii + 209 p.
157. COOK, GILLIAN P. *Spatial Dynamics of Business Growth in the Witwatersrand.* 1975. x + 144 p.
160. MEYER, JUDITH W. *Diffusion of an American Montessori Education.* 1975. xi + 97 p.
162. LAMB, RICHARD F. *Metropolitan Impacts on Rural America.* 1975. xii + 196 p.
163. FEDOR, THOMAS STANLEY. *Patterns of Urban Growth in the Russian Empire during the Nineteenth Century.* 1975. xxv + 245 p.
164. HARRIS, CHAUNCY D. *Guide to Geographical Bibliographies and Reference Works in Russian or on the Soviet Union.* 1975. xviii + 478 p.
165. JONES, DONALD W. *Migration and Urban Unemployment in Dualistic Economic Development.* 1975. x + 174 p.
166. BEDNARZ, ROBERT S. *The Effect of Air Pollution on Property Value in Chicago.* 1975. viii + 111 p.
167. HANNEMANN, MANFRED. *The Diffusion of the Reformation in Southwestern Germany, 1518-1534.* 1975. ix + 235 p.
168. SUBLETT, MICHAEL D. *Farmers on the Road: Interfarm Migration and the Farming of Noncontiguous Lands in Three Midwestern Townships. 1939-1969.* 1975. xiii + 214 p.
169. STETZER, DONALD FOSTER. *Special Districts in Cook County: Toward a Geography of Local Government.* 1975. xi + 177 p.
171. SPODEK, HOWARD. *Urban-Rural Integration in Regional Development: A Case Study of Saurashtra, India—1800-1960.* 1976. xi + 144 p.
172. COHEN, YEHOSHUA S., and BRIAN J. L. BERRY. *Spatial Components of Manufacturing Change.* 1975. vi + 262 p.
173. HAYES, CHARLES R. *The Dispersed City: The Case of Piedmont, North Carolina.* 1976. ix + 157 p.
174. CARGO, DOUGLAS B. *Solid Wastes: Factors Influencing Generation Rates.* 1977. 100 p.
176. MORGAN, DAVID J. *Patterns of Population Distribution: A Residential Preference Model and Its Dynamic.* 1978. xiii + 200 p.
177. STOKES, HOUSTON H.; DONALD W. JONES; and HUGH M. NEUBURGER. *Unemployment and Adjustment in the Labor Market: A Comparison between the Regional and National Responses.* 1975. ix + 125 p.
180. CARR, CLAUDIA J. *Pastoralism in Crisis. The Dasanetch and Their Ethiopian Lands.* 1977. xx + 319 p.
181. GOODWIN, GARY C. *Cherokees in Transition: A Study of Changing Culture and Environment Prior to 1775.* 1977. ix + 207 p.
183. HAIGH, MARTIN J. *The Evolution of Slopes on Artificial Landforms, Blaenavon, U.K.* 1978. xiv + 293 p.
184. FINK, L. DEE. *Listening to the Learner: An Exploratory Study of Personal Meaning in College Geography Courses.* 1977. ix + 186 p.
185. HELGREN, DAVID M. *Rivers of Diamonds: An Alluvial History of the Lower Vaal Basin, South Africa.* 1979. xix + 389 p.

186. BUTZER, KARL W., ed. *Dimensions of Human Geography: Essays on Some Familiar and Neglected Themes.* 1978. vii + 190 p.

187. MITSUHASHI, SETSUKO. *Japanese Commodity Flows.* 1978. x + 172 p.

188. CARIS, SUSAN L. *Community Attitudes toward Pollution.* 1978. xii + 211 p.

189. REES, PHILIP M. *Residential Patterns in American Cities: 1960.* 1979. xvi + 405 p.

190. KANNE, EDWARD A. *Fresh Food for Nicosia.* 1979. x + 106 p.

192. KIRCHNER, JOHN A. *Sugar and Seasonal Labor Migration: The Case of Tucumán, Argentina.* 1980. xii + 174 p.

194. HARRIS, CHAUNCY D. *Annotated World List of Selected Current Geographical Serials, Fourth Edition. 1980.* 1980. iv + 165 p.

196. LEUNG, CHI-KEUNG, and NORTON S. GINSBURG, eds. *China: Urbanizations and National Development.* 1980. ix + 283 p.

197. DAICHES, SOL. *People in Distress: A Geographical Perspective on Psychological Well-being.* 1981. xiv + 199 p.

198. JOHNSON, JOSEPH T. *Location and Trade Theory: Industrial Location, Comparative Advantage, and the Geographic Pattern of Production in the United States.* 1981. xi + 107 p.

199-200. STEVENSON, ARTHUR J. *The New York–Newark Air Freight System.* 1982. xvi + 440 p.

201. LICATE, JACK A. *Creation of a Mexican Landscape: Territorial Organization and Settlement in the Eastern Puebla Basin, 1520-1605.* 1981. x + 143 p.

202. RUDZITIS, GUNDARS. *Residential Location Determinants of the Older Population.* 1982. x + 117 p.

203. LIANG, ERNEST P. *China: Railways and Agricultural Development, 1875-1935.* 1982. xi + 186 p.

204. DAHMANN, DONALD C. *Locals and Cosmopolitans: Patterns of Spatial Mobility during the Transition from Youth to Early Adulthood.* 1982. xiii + 146 p.

206. HARRIS, CHAUNCY D. *Bibliography of Geography. Part II: Regional. Volume 1. The United States of America.* 1984. viii + 178 p.

.207-208. WHEATLEY, PAUL. *Nagara and Commandery: Origins of the Southeast Asian Urban Traditions.* 1983. xv + 472 p.

209. SAARINEN, THOMAS F.; DAVID SEAMON; and JAMES L. SELL, eds. *Environmental Perception and Behavior: An Inventory and Prospect.* 1984. x + 263 p.

210. WESCOAT, JAMES L., JR. *Integrated Water Development: Water Use and Conservation Practice in Western Colorado.* 1984. xi + 239 p.

211. DEMKO, GEORGE J., and ROLAND J. FUCHS, eds. *Geographical Studies on the Soviet Union: Essays in Honor of Chauncy D. Harris.* 1984. vii + 294 p.

212. HOLMES, ROLAND C. *Irrigation in Southern Peru: The Chili Basin.* 1986. ix + 199 p.

213. EDMONDS, RICHARD LOUIS. *Northern Frontiers of Qing China and Tokugawa Japan: A Comparative Study of Frontier Policy.* 1985. xi + 209 p.

214. FREEMAN, DONALD B., and GLEN B. NORCLIFFE. *Rural Enterprise in Kenya: Development and Spatial Organization of the Nonfarm Sector.* 1985. xiv + 180 p.

215. COHEN, YEHOSHUA S., and AMNON SHINAR. *Neighborhoods and Friendship Networks: A Study of Three Residential Neighborhoods in Jerusalem.* 1985. ix + 137 p.

216. OBERMEYER, NANCY J. *Bureaucrats, Clients, and Geography: The Bailly Nuclear Power Plant Battle in Northern Indiana.* 1989. x + 135 p.

217-218. CONZEN, MICHAEL P., ed. *World Patterns of Modern Urban Change: Essays in Honor of Chauncy D. Harris.* 1986. x + 479 p.

219. KOMOGUCHI, YOSHIMI. *Agricultural Systems in the Tamil Nadu: A Case Study of Peruvalanallur Village.* 1986. xvi + 175 p.

220. GINSBURG, NORTON; JAMES OSBORN; and GRANT BLANK. *Geographic Perspectives on the Wealth of Nations.* 1986. ix + 133 p.

221. BAYLSON, JOSHUA C. *Territorial Allocation by Imperial Rivalry: The Human Legacy in the Near East.* 1987. xi + 138 p.

222. DORN, MARILYN APRIL. *The Administrative Partitioning of Costa Rica: Politics and Planners in the 1970s.* 1989. xi + 126 p.

223. ASTROTH, JOSEPH H., JR. *Understanding Peasant Agriculture: An Integrated Land-Use Model for the Punjab.* 1990. xiii + 173 p.

224. PLATT, RUTHERFORD H.; SHEILA G. PELCZARSKI; and BARBARA K. BURBANK, eds. *Cities on the Beach: Management Issues of Developed Coastal Barriers.* 1987. vii + 324 p.

225. LATZ, GIL. *Agricultural Development in Japan: The Land Improvement District in Concept and Practice.* 1989. viii + 135 p.

226. GRITZNER, JEFFREY A. *The West African Sahel: Human Agency and Environmental Change.* 1988. xii + 170 p.

227. MURPHY, ALEXANDER B. *The Regional Dynamics of Language Differentiation in Belgium: A Study in Cultural-Political Geography.* 1988. xiii + 249 p.

228-229. BISHOP, BARRY C. *Karnali under Stress: Livelihood Strategies and Seasonal Rhythms in a Changing Nepal Himalaya.* 1990. xviii + 460 p.

230. MUELLER-WILLE, CHRISTOPHER. *Natural Landscape Amenities and Suburban Growth: Metropolitan Chicago, 1970-1980.* 1990. xi + 153 p.

231. WILKINSON, M. JUSTIN. *Paleoenvironments in the Namib Desert: The Lower Tumas Basin in the Late Cenozoic.* 1990. xv + 196 p.

232. DUBOIS, RANDOM. *Soil Erosion in a Coastal River Basin: A Case Study from the Philippines.* 1990. xii + 138 p.